Frontiers in African Business Research

Series editor

Almas Heshmati, Jönköping International Business School,
Jönköping, Sweden

This book series publishes monographs and edited volumes devoted to studies on entrepreneurship, innovation, as well as business development and management-related issues in Africa. Volumes cover in-depth analyses of individual countries, regions, cases, and comparative studies. They include both a specific and a general focus on the latest advances of the various aspects of entrepreneurship, innovation, business development, management and the policies that set the business environment. It provides a platform for researchers globally to carry out rigorous analyses, to promote, share, and discuss issues, findings and perspectives in various areas of business development, management, finance, human resources, technology, and the implementation of policies and strategies of the African continent. Frontiers in African Business Research allows for a deeper appreciation of the various issues around African business development with high quality and peer reviewed contributions. Volumes published in the series are important reading for academicians, consultants, business professionals, entrepreneurs, managers, as well as policy makers, interested in the private sector development of the African continent.

More information about this series at http://www.springer.com/series/13889

Nnaemeka Vincent Emodi

Energy Policies for Sustainable Development Strategies

The Case of Nigeria

Springer

Nnaemeka Vincent Emodi
College of Business, Law and Governance
James Cook University
Cairns, QLD
Australia

ISSN 2367-1033 ISSN 2367-1041 (electronic)
Frontiers in African Business Research
ISBN 978-981-10-0973-0 ISBN 978-981-10-0974-7 (eBook)
DOI 10.1007/978-981-10-0974-7

Library of Congress Control Number: 2016939040

Printed on acid-free paper

This Springer imprint is published by Springer Nature
The registered company is Springer Science+Business Media Singapore Pte Ltd.

Contents

Abstract

This research developed the long-term energy scenarios for Nigeria considering the impact of the vital factors that may influence energy polices in the future energy system. The energy scenarios were developed through the application of the Long-Range Energy Alternatives Planning System (LEAP) model. The Nigerian LEAP model was developed in order to identify the future energy demand and how it could be met using a least-cost combination of technology options without similar expansion in greenhouse gases. The model developed incorporated four policy scenarios which differ from each other and this was intended to capture the vital factors that may influence energy policies in the future.

The factors taken as parameters included GDP, households, population and urbanization growth rates, and growth rates of energy-intensive sectors. The four scenarios developed are as follows: the reference (Ref), low carbon moderate (LCM), low carbon advance (LCA), and green optimistic (GO) scenarios. The results of the modeled scenarios showed that energy demand is expected to grow at an annual growth rate of 3.58 % (Ref), 3.53 % (LCM), 2.95 % (LCA), and 2.61 % (GO). The Ref scenario energy demand was the highest with 3,075PJ by 2040, while the GO scenario was the lowest at 2,249.2PJ. The GHG emission rate was very low for GO scenario (124.4 MMTCDE) as compared to other scenarios and this was due to the high level of renewables and energy efficiency application into the energy mix.

The level of energy policies such as various degrees of energy efficiency, fuel and technology switching were increased from the LCM scenario, which had a moderate policy implementation, the advance LCA scenario, and the more aggressive GO scenario. Furthermore, a cost-benefit analysis was carried out in order to ascertain the cost of implementing some policies and strategies in Nigeria which includes energy efficiency and fuel/technology switching. The results showed that it would cost Nigeria 1.69 billion US dollars to implement policies in the LCM, 23.8 billion US dollars for the LCA, and 41.4 billion US dollars in the GO scenario.

The least-cost electricity generation for the power plants proposed in the scenarios showed that among the power plants, on-shore wind power and small hydropower provided the overall least cost of electricity generation. For fossil fuel power plants, CCGT was identified to be the least option for electricity generation, as well as the lowest GHG emitting power plants besides biomass which was considered a low carbon technology. From the results in general, it was observed that low carbon and renewable technologies will have an important role to play in the realization of a low carbon development in Nigeria.

To achieve this feat, this study further explored some strategies that can ensure a low carbon development in Nigeria with a view of attaining green growth. These strategies includes thinking and adopting the green growth ideology, energy policy reforms, long-term energy planning and target, energy regulation and standards, environmental tax reforms, urban planning, efficient building design, efficiency of the energy and transport system.

This research presents some significance in the sense that it applied a bottom-up approach for the Nigerian energy model, presents a cost-benefit analysis and least-cost electricity generation options, and finally presenting some strategic energy policies. The findings from this study can be used as a guide in the development of energy policies and sustainable strategies for the attainment of a low carbon development on the long term in Nigeria.

Keywords Nigeria LEAP model · Scenario analysis · Least-cost electricity generation · Cost-benefit analysis · Sustainable strategies · Low carbon development

List of Figures

List of Tables

Chapter 1
Overview

Abstract This chapter introduces the general overview of the study. It starts with an introduction which presents a general overview of energy, environment, society, and the need for energy planning. The background problem describes the existing problems and the need for the research, which led to the statement of the research problem. The research questions are set in line with the statement of the research problems, while the research objectives intend to solve them. The methodology used in this study, its scope, and significance are all presented in this chapter.

Keywords Energy supply and consumption · Energy commission of Nigeria · Energy planning · LEAP model

1.1 Introduction

During the last century, the world experienced a transition in energy use from coal based to petroleum based. Due to the increase in globalization and industrialization, the global demand for energy has increased exponentially (Suganthi and Samuel 2012). With the increase in demand for energy, its production has also increased such that about 80 % of global energy supply comes from fossil fuel (IEA 2008). The increase in the reliance on fossil fuel has resulted in an increase in global greenhouse gas emissions and this has raised issues about the sustainability of the environment due to problems such as climate change and depletion of resources (Van Ruijvan et al. 2008). Besides the high demand of energy by industrialized nations, developing countries have rapidly increased their energy consumption which has contributed to sustainability problems (Urban et al. 2007). Therefore, the acknowledgement of the increasing influence of developing countries on the global energy settings cannot be overemphasized.

Energy consumption has been one of the most reliable indicators of the development and quality of life reached by a country, and the necessity of satisfying a forecasted energy demand for a given period of time is the rationale for energy

N.V. Emodi, *Energy Policies for Sustainable Development Strategies*,
Frontiers in African Business Research, DOI 10.1007/978-981-10-0974-7_1

planning (Cormio et al. 2003). As defined by the World Energy Council, "Energy planning is that part of economics applied to energy problems, taking into account the analysis of energy supply and demand, as well as implementation of the means for ensuring coverage of energy needs in a national or international context" (Dictionary 1992). The methods used in energy planning are classified into planning models, by analogy, and by inquiry (Kleinpeter 1995).

Consistency in the methods depends on the time interval which can be short-term (about 5–10 years), medium-term (10–20 years) and long-term (above 20 years). Since the mid-1970s, energy modeling has been a tool for national energy planning and was used to understand the implications of the first oil crisis (Nakata 2004). Researchers since then have employed different energy models to address policy and planning concerns of energy, economy, and the environment (Pandey 2002).

1.2 Background of the Problem

In many developing countries, mostly in Africa and some Southeast Asia countries, access to clean energy is known to be unreliable and at high disruption costs which affects production efficiency and competitiveness (Emodi and Yusuf 2015). Being endowed with the widest possible range of energy resources, the African continent has experienced a relatively low energy consumption in general and electricity consumption in particular (Moyo 2012).

Nigeria is experiencing a remarkable paradox—the abundance of energy resources and widespread energy poverty with about 40 % of the population having access to the grid electricity supply, while 70 % depends on firewood (Eleri et al. 2012). This depends on firewood which constitutes a major indoor pollution hazard and has caused the death of about 79,000 Nigerians due to smoke inhalation in 2011 (Eleri and Onuvae 2011). According to a study by the World Health Organization in 2013, the death caused by smoke inhalation from firewood use by women reached 98,000 (Emodi and Boo 2015a).

The Nigerian Government's action to address the issue of energy poverty has been through the increase in the amount of gas power plants for electricity generation (Emodi and Boo 2015b), while plans are still in progress to introduce other source of energy services. The plans are being set up by the government agency, the Energy Commission of Nigeria (ECN), which is responsible for the strategic planning and coordination of national policies in the field of energy in all its various forms in Nigeria (ECN 2015).

The ECN carried out a study to ascertain the projected future energy demand and supply under four scenarios. This study was carried out using the Model for the Energy Supply Strategy Alternatives and their General Environmental Impact (MESSAGE) and the results are included in the current National Energy Master Plan (NEMP 2014).

1.3 Statement of the Problem

The statements of the problem in this study are presented as follows:

(1) The study carried out by the ECN did not develop scenarios considering various parameters which can influence energy demand. The only parameter considered was economic growth rate (GDP) which is not the only determinant of increase in energy demand. Other variables such as population, income, household size, technology, energy prices, etc. also influence energy demand. Also, least cost options were not observed in the ECN study;
(2) The impact of future energy policy implementation and strategies were not considered in the ECN study. The impact of various energy policies is expected to alter the energy consumption pattern of a country on the short, medium, and long term. Also incentives such as Feed-in-Tariff may also affect energy consumption pattern;
(3) The effects of greenhouse gases (GHGs) on the environment were not taken into account in the ECN study. It is worthy to note that the contribution of GHGs to climate change cannot be left out when carrying out energy modeling because this has a direct impact on the society. The effect of climate change can be observed in almost all parts of Nigeria and as such, the consideration for GHGs reduction is important.
(4) Provisions were not made to identify power plants with least-cost electricity generation potential in Nigeria. It is worthy to note that the contribution is very important because it aids in the identification of the most economical generation expansion plan to achieve a certain level of reliability to meet the forecasted electricity demand. Furthermore, investment on energy technologies can be decided based on cheaper power plants and a considerable lower emission rate.

1.4 Research Questions

This study presents two research questions which were derived from the research problems. The first question was developed from the first, third, and last research problems stated in Sect. 1.2, while the second research question is concerned about the problem, also stated in Sect. 1.2. The research questions in this study are presented as follows:

(1) What are the projected energy demands in Nigeria and how will the demand be met using a least-cost combination of technology options without a similar expansion in GHGs?
(2) Which sustainable energy policy options are recommended to improve the socio-economic and energy situation in Nigeria?

1.5 Research Objectives

In order to answer the research questions and solve the research problems, this study has the following objectives:

(1) to develop the long-term energy scenarios for Nigeria using the Long-Range Energy Alternatives Planning System (LEAP), considering vital factors which may influence energy policies in the future;
(2) to examines the long-term energy impact of the scenarios in terms of energy supply mix, energy demand, electricity generation mix, and GHG emissions;
(3) to further explore the least-cost electricity generation options;
(4) to recommend energy policy options that can improve the energy situation in Nigeria using a low carbon approach that is sustainable in the long run.

1.6 Methodology

A graphical representation of the methodology is presented in Fig. 1.1, and this shows the general overview of methodologies and procedures employed in this research. It includes the research problems, questions, objectives, and methodological framework. Furthermore, included in this section is a detailed step-by-step methodological procedure.

The research begins with a general overview of the Nigerian energy sector with an extensive review of the various energy resources, past trend of energy demand

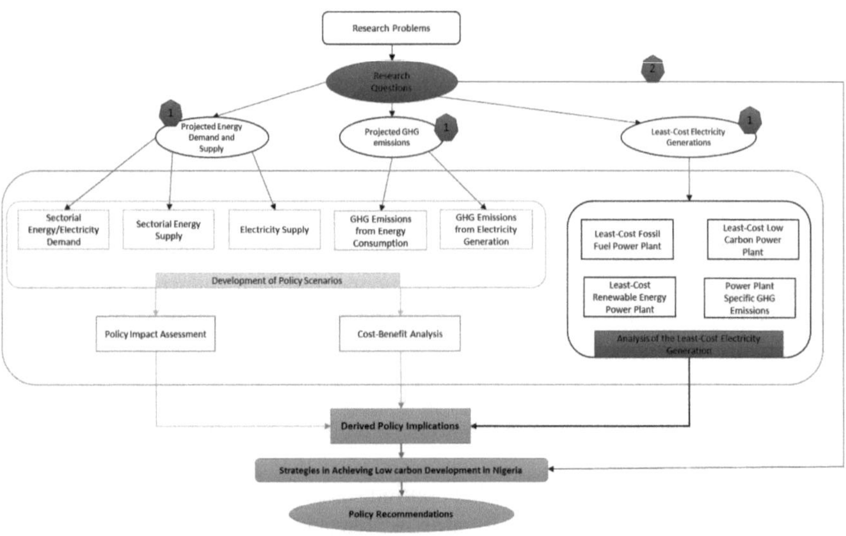

Fig. 1.1 Methodologies

and supply, and an insight into the roles of government agencies and policies. This is intended to guide the reader on the current situation of the energy sector in Nigeria as it relates to this study. This is followed by a review on the existing energy forecasting models and literature applying the LEAP model.

This was done so as to explore the characteristic features of energy forecasting models and to establish a basis for the selection of the model used in this study. The literature review examines the scholarly contributions in the economic literature which applied the LEAP model on various economic sectors in different countries of the world. This further establishes the rationale for the development of four policy scenarios which includes one reference scenario and three alternative scenarios for the Nigerian LEAP model.

The developed policy scenarios were projected from the base year which was 2010, to the target year which is 2040. This was done considering various parameters that are capable of altering the energy system in the future and they include GDP, income, population, households size, and sectorial growth rate. The projection/forecast includes energy demand and supply, electricity demand and supply (including capacity expansion), GHG emissions, and cost-benefit analysis.

The developed alternative scenarios were examined to assess the impact of various policies and strategies applied to each scenario in comparison to the reference scenario. An extension of the methodology includes the least-cost electricity generation which was carried out through optimization of the selected power plants used in the scenarios' analysis. The power plants specific GHGs emissions over the period under observation were also carried out using the least-cost electricity generation methodology.

Sustainable strategies were developed from various successful country case studies and this was in relation to the developed scenarios. Policy implications were then drawn out of the scenarios results, cost-benefits analysis, and least-cost electricity generation to provide policy recommendations that are sustainable in the long run.

1.7 Scope of Research

This study will focus on modeling the future energy demand and supply in Nigeria using the LEAP model. A graphical representation of the research scope is shown in Fig. 1.2. Emphasis is laid on (i) primary energy supply mix which includes coal, hydro, natural gas, crude oil, biomass, and other renewable resources. (ii) Transformation; biofuel, charcoal, compressed natural gas, diesel, electricity, firewood, kerosene, liquefied petroleum gas, and petrol. (iii) Final energy demand; household, industry, agriculture, and transport sectors (iv) GHG emission reductions potentials (v) least-cost electricity generation.

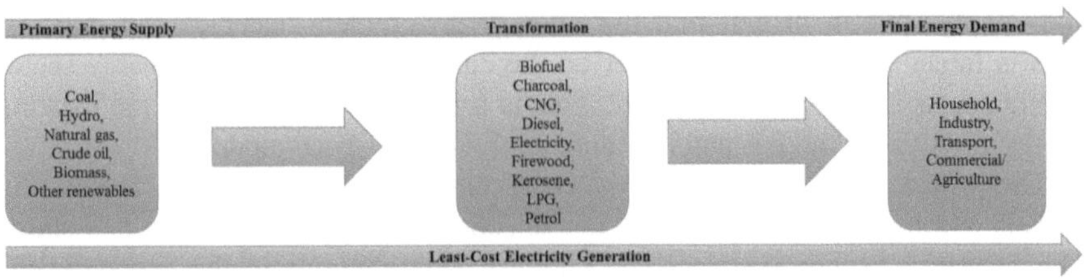

Fig. 1.2 Scope of research

The LEAP modeling scenarios: the Reference (Ref), Low Carbon Moderate (LCM), Low Carbon Advance (LCA), and Green Optimistic (GO) scenarios. The scenarios which include various energy technologies, services, strategies, and energy policies were analyzed for the most appropriate low carbon approach of meeting energy demand at a least cost and low GHG emission.

1.8 Significance of the Research

The major significance of this research is:

- The Nigerian LEAP model will present a bottom-up energy model for the Nigerian energy system. The bottom-up model is the first of its kind in the Nigerian context and covers important energy-intensive sectors.
- The three distinctive alternative scenarios expand the possibility of capturing the future energy pathways for Nigeria. Since the future is uncertain, employing more changes in each scenario increases the chance of a possible future pathway for the Nigerian energy system.
- This study will add to the body of knowledge in the Nigerian context by exploring the influences of energy policies and strategies on the future energy services in Nigeria. This was made possible through the policy impact assessment, cost-benefit analysis, and the least-cost electricity generation. This is important to the Nigerian government and policymakers in particular, and the society in General.

1.9 Structure of the Study

This study is divided into seven chapters and graphical representation of the study structure is shown in Fig. 1.3. Chapter 2 presents the current situation of the energy sector in Nigeria. This chapter extensively reviews the energy resource potentials in Nigeria, energy supply and consumption, and further presents an insight into the existing government agencies and policies concerned with energy

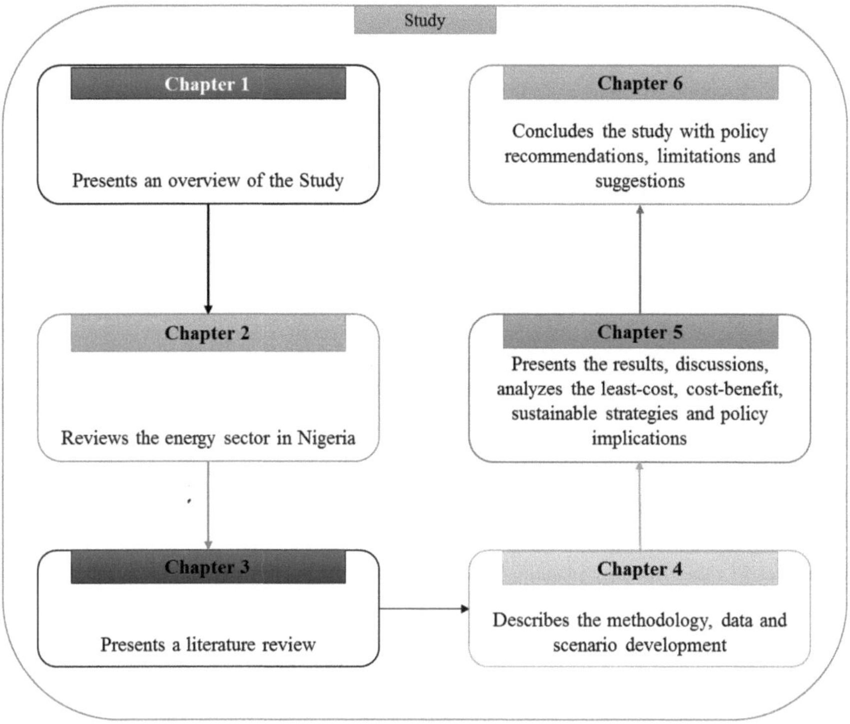

Fig. 1.3 Structure of the study

development. Chapter 3 presents a literature review on both existing energy forecasting models and literature that applied the LEAP model. The chapter intends to explain the characteristics of energy models and further reviews available literature with the LEAP model application. Chapter 4 describes the methodological approach applied in this study.

This includes the LEAP model description and the algorithm used in the forecasting of future energy demand, supply, GHG emissions, cost-benefit analysis, and least-cost electricity generation. The scenario development and data collection process are also discussed in Chap. 4.

The results of the developed policy scenarios are presented in Chap. 5. This chapter also contains the discussion on the policy impact on the scenarios, the cost-benefit analysis, least-cost electricity generation, and. presents some sustainable strategies for low carbon development in Nigeria. Further, the chapter is summarized with some policy implications. Chapter 6 concludes this study with some policy recommendations, limitations, and suggestions for further studies.

References

Cormio C, Dicorato M, Minoia A, Trovato M (2003) A regional energy planning methodology including renewable energy sources and environmental constraints. Renew Sustain Energy Rev 7(2):99–130

Dictionary E (1992) World Energy Council. Jouve SI, Paris

Eleri EO, Onuvae P (2011) Low-carbon Africa: leapfrogging to green future: low carbon Nigeria. www.christainaid.org.uk/resources/policy/climate/low-carbon-africa.aspx. Accessed 18 July 2015

Eleri EO, Ugwu O, Onuvae P (2012) Expanding access to pro-poor energy services in Nigeria. International Center for Energy, Environment and Development, Abuja, Nigeria

Emodi NV, Boo KJ (2015a) Sustainable energy development in Nigeria: current status and policy options. Renew Sustain Energy Rev 51:356–381

Emodi NV, Boo KJ (2015b) Sustainable energy development in Nigeria: overcoming energy poverty. Int J Energy Econ Policy 5(2):580–597

Emodi NV, Yusuf SD (2015) Improving electricity access in Nigeria: obstacles and the way forward. Int J Energy Econ Policy 5(1):335–351

Energy Commission of Nigeria (ECN) (2015). www.energy.gov.ng. Accessed 6 Aug 2015

International Energy Agency (IEA) (2008) Key world energy statistics. Technical report, International Energy Agency

Kleinpeter M (1995) Energy planning and policy. In: Fuel and energy abstracts, vol 5, No 36, p 382

Moyo B (2012) Do power cuts affect productivity? A case study of Nigerian manufacturing firms. Int Bus Econ Res J (IBER) 11(10):1163–1174

Nakata T (2004) Energy-economic models and the environment. Prog Energy Combust Sci 30(4):417–475

National Energy Master Plan (NEMP) (Draft Revised Edition) (2014) Energy Commission of Nigeria (ECN). www.energy.gov.ng. Accessed 20 July 2015

Pandey R (2002) Energy policy modelling: agenda for developing countries. Energy Policy 30(2):97–106

Suganthi L, Samuel AA (2012) Energy models for demand forecasting—a review. Renew Sustain Energy Rev 16(2):1223–1240

Urban FRMJ, Benders RMJ, Moll HC (2007) Modelling energy systems for developing countries. Energy Policy 35(6):3473–3482

Van Ruijven B, Urban F, Benders RM, Moll HC, Van Der Sluijs JP, De Vries B, Van Vuuren DP (2008) Modeling energy and development: an evaluation of models and concepts. World Dev 36(12):2801–2821

Chapter 2
The Energy Sector in Nigeria

Abstract The chapter presents the status quo of Nigeria's energy sector and is divided into five sections. The first extensively reviews the various conventional and renewable energy resources in Nigeria. The second and third sections presents the primary energy supply and consumption, respectively. The forth section presents an insight into the various government ministries, parastatals, and agencies that are relevant in the Nigerian energy sector. The last section in this chapter explores the Nigerian energy policies and strategies.

Keywords Nigerian power sector · Conventional energy resources · Renewable energy · Nigerian energy policies · Government ministries and agencies

2.1 A Brief History on the Nigerian Power Sector

The Federal Government of Nigeria (FGN) has been responsible for the formulation of energy policy development and regulation, including operation and investment in the Nigerian energy sector before 2005,[1] after which the FGN established the Electricity Power Sector Reform Act (Federal Government of Nigeria [FGN] 2005). The Federal Ministry of Power (FMP) conducted regulation of the power sector, while the National Electric Power Authority (NEPA) handled operation in the sector. Part of the responsibility of NEPA was power generation, transmission, and distribution.[2] NEPA, however, ran a monopoly system from its inception in 1972 to its defunct in 2005, with control of power generation capacity of about 94 %, while transmission, system operators, distribution, and their marketing sector was 100 % owned by them (Nigeria Bureau of Public Enterprise [NBPE] 2015).

[1]www.energypedia.info/wiki/Nigeria_Energy_Situation.

[2]www.nigeriaelectricityprivatisation.com/?page_id=2.

N.V. Emodi, *Energy Policies for Sustainable Development Strategies*,
Frontiers in African Business Research, DOI 10.1007/978-981-10-0974-7_2

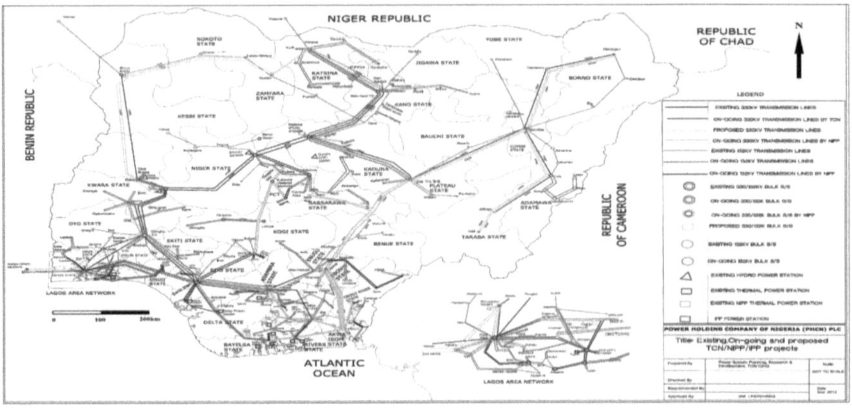

Fig. 2.1 Map of existing, on-going, and proposed TCN/NIPP/IPP projects for electricity transmission in Nigeria. *Source* TCN (2015), Akpan (2015)

Inefficiencies in operations and financial performance in NEPA led the amendment of the Electricity and NEPA Acts by the FGN in 1998 to remove the monopoly held by NEPA and encouraged the participation of the private sector.[3] A reform agenda was specified in the National Electric Power Policy in 2001, while the legal basis for the unbundling of NEPA and formulation of successor companies (including privatization) was provided in the EPSRA (National Mirror 2014). NEPA was restructured to form the Power Holding Company of Nigeria (PHCN) in 2007 which stopped existing after September 30th 2013.[4]

The FGN sorted various means to sell off its stake in the electricity services industry[5] (i.e. privatization), but retained the transmission grid as a public entity (Aladejare 2014). The generation companies are now called the GENCOs, while the distribution companies are called the DISCOs, and the FGN operates the Transmission Company of Nigeria[6] (TCN) (KPMG 2013). The generators and transmission lines are interconnected in the national grid system which is controlled at the National Control Center, Oshogbo. Figure 2.1 shows the TCN/NIPP[7]/IPP[8] projects in Nigeria as presented by PHCN.[9]

[3]www.placng.org/new/laws/E7.pdf.

[4]www.nigeriaelectricityprivatisation.com/.

[5]The state-owned electricity generation companies was placed for sale by the Nigerian government in two ways in which one was the outright sale of the thermal power stations, while the other was through concession of the hydroelectric stations.

[6]The TCN is divided into two divisions; the systems operator division and a market operator division.

[7]National Integrated Power Project.

[8]Independent Power Producer.

[9]This was presented in 2012.

The FGN took the next step in setting up the Nigerian Electricity Regulatory Commission (NERC) and the Nigerian Bulk Electricity Trading Plc. (NBET).[10] The Operators of the Nigerian Electricity Market[11] (ONEM) was established with the responsibility of the wholesale market and settlement operator. After the establishment of the agencies, the FGN then placed the new NIPP power plants for sale. Although most of the GENCOs companies were bound to suffer a loss, the FGN allocated N50 billion to them so as to reduce the impact of the loss (KPMG 2013). The NBET buys electric energy generated from the GENCOs and sells it to the DISCOs for sale to the final electricity consumers.[12]

The steps discussed above were taken to unbundle the PHCN and this is classified as the pre-transition electricity market (pre-TEM) development.[13] However, until the TEM becomes fully functional, some rules govern the electricity market in which the GENCOs charge the ONEM (Detail Commercial Solicitors 2014). The financing requirements for the market are set by the ONEM on the basis of the MYTO 11.[14] The aim was to develop a framework which can oversee the arrangement of electricity trading during the pre-TEM period and is still in effect. The TEM will be functional when the electricity market becomes fully privatized and private sector oriented.[15] The location of the GENCOs (power plants) is shown in Fig. 2.2.

2.2 The Nigerian Energy Resources

Several energy resources are available in Nigeria in abundant proportions. This includes conventional and non-conventional (renewable) energy resources, and are vastly distributed across the regions of the country. The energy resources in Nigeria are classified as conventional and renewable energy resources, and they are discussed as follows.

[10]The NBET was set-up, although not fully effective, but was intended to come into full operation when the Nigerian electricity market becomes completely privatized and then the power purchase agreements will be signed and passed on to the DISCOs (The Presidency 2013).

[11]This responsibility extends to the management of the metering system of the TCN, DISCOs and the GENCOs.

[12]See www.nbet.com.ng.

[13]For a more insight into this, see Detail Commercial Solicitors (2015).

[14]Multi-Year Tariff Order 2.

[15]See www.lexology.com/library/detail.aspx?g=f77e24d6-8338-47d6-9e4d-ed5b89a51862.

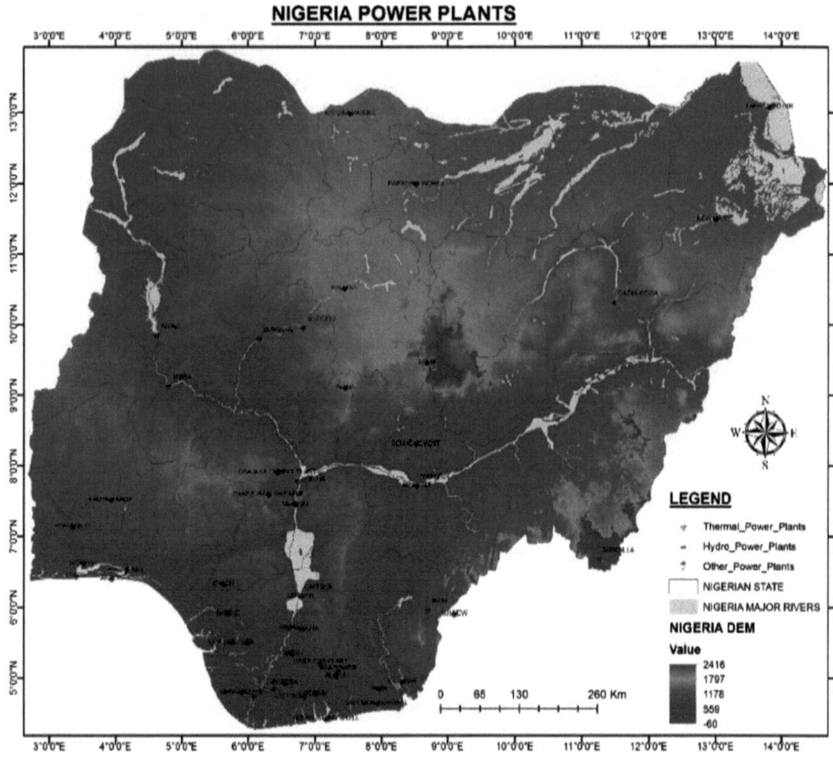

Fig. 2.2 Nigeria Power Plants. *Source* Nwoko (2015a)

2.2.1 Conventional Energy Resources

Nigeria has considerable reserves of conventional energy resources. It is one of the world's largest producers of oil and it has the largest reserves of natural gas in the African Continent. It therefore became the world's fourth leading exporter of liquefied natural gas (LNG) in 2012. Nigeria is also a member of the Organization of the Petroleum Exporting Countries (OPEC), which it joined in 1971 after over 10 years of oil production that began in the late 1950s (EIA 2014a). Coal reserves stand at 2.175 billion tons, but production has long since ceased (in the 1950s) as the government has concentrated on the oil and gas resources.

Nigeria is also rich in tar sand or oil sand, which is a combination of clay, sand, water, and bitumen (a heavy black viscous oil). Tar sands can be mined and processed to extract the oil-rich bitumen, which can be refined into oil (Oil Shale and Tar Sands Programmatic (EIS 2014). Table 2.1 presents the conventional energy reserves in Nigeria and their potentials.

According to the US Energy Information Agency estimate, the total primary energy consumption in Nigeria in 2012 was about 4.5 quadrillion Btu (British

Table 2.1 Conventional energy reserves in Nigeria and their potentials

Resource type	Reserves		Production	Domestic Utilization (natural units)
	Natural units	Energy units (Btoe)		
Natural gas	187 trillion SCF	4.19	6 billion SCF/day	3.4 billion SCF/day
Crude oil	36.22 billion barrels	5.03	2.5 million barrels/day	450,000 barrels/day
Tar sands	31 billion barrels of equivalent	4.31	Insignificant	Insignificant
Coal and lignite	2.175 billion ton	1.52	–	–
Nuclear element	None	–	–	–

Source National Bureau of Statistics [NBS] (2007)

thermal units). This comprised 80 % from traditional biomass and waste (wood, charcoal, manure, and crop residue) and much smaller percentages from oil and natural gas. The Nigerian oil and gas sector is regulated by the Nigerian National Petroleum Corporation (NNPC), which was established in 1977 with the secondary responsibility of overseeing the development of the upstream and downstream oil sectors.[16]

Despite the large energy resources in Nigeria, energy consumption is relatively low compared with other African countries with comparable energy resources (Fig. 2.3). This low energy consumption is due to the recurrent scarcity of petroleum products at vehicle petrol stations, while frequent electricity "black-outs" have resulted in a high reliance by the Nigerian populace on personal electricity generators.

Despite the scarcity of petroleum products, energy demand has been increasing in Nigeria, because of the increase in economic development and the population growth. According to Sambo et al. (2006), the major driver behind increasing energy demand is the population growth, while the most important determinant is the level of economic activity, measured by the country's gross domestic product (GDP).

Nigeria's population is projected to grow from 178,516,904 (as of 2014) to 440,355,062 by 2050 (Fig. 2.4). To address the needs of this increasing population, the Energy Commission of Nigeria (ECN) analyzed the country's energy sector from 2000 to 2030 using the Wien Automatic System Planning (WASP) package and the Model for Analysis of Energy Demand (MAED).

[16]See "Oil and Gas in Nigeria". Available online at: www.mbendi.com/indy/oilg/af/ng/p0005.htm.

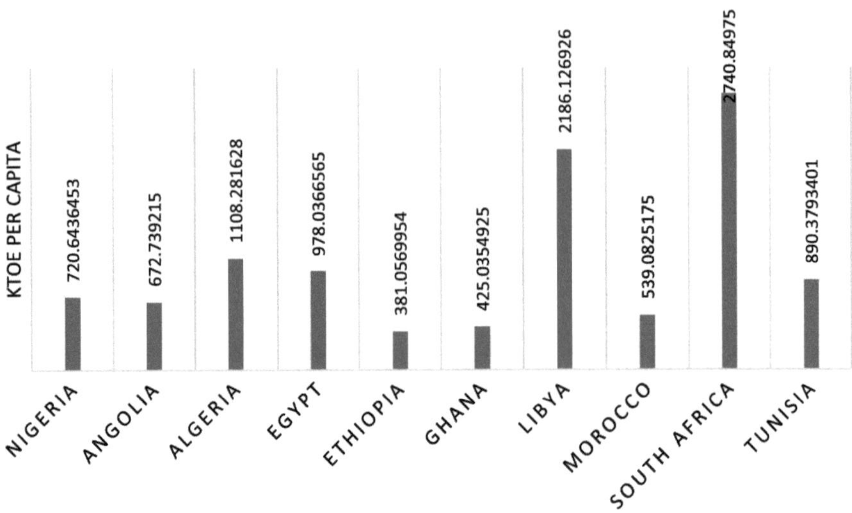

Fig. 2.3 Energy consumption per capita in some African countries in 2012 (The World Bank 2014)

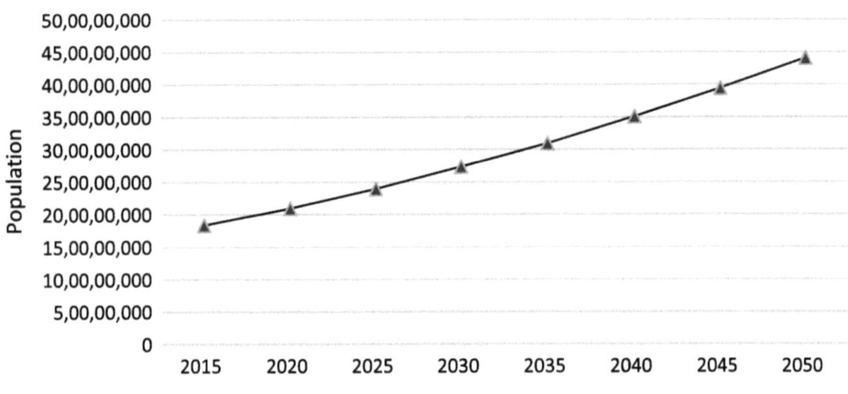

Fig. 2.4 Projections of population growth in Nigeria (World Meters 2015)

The results (Fig. 2.5), based on reference, high growth, and two optimistic (11.5 and 13 % GDP growth) scenarios, project that energy demand in Nigeria will increase by 2.5, 3, 3.5, and 4.5 times, respectively, from 2000 to 2015, and by 8, 13, 17, and 22.5 times, respectively, from 2000 to 2030.

According to ECN, the increase in energy demand will develop in line with the high level of economic activity expected in Nigeria, as measured by the total GDP. The sectorial energy demand (Fig. 2.6) shows that although the industrial sector had lower energy demand in 2005 (8.05 Million Tons of Oil Equivalent (Mtoe)), it will have the highest energy demand by 2030 at 145.21 Mtoe with GDP growth rate of 16.27 %, as the economy begins to improve based on increased industrial

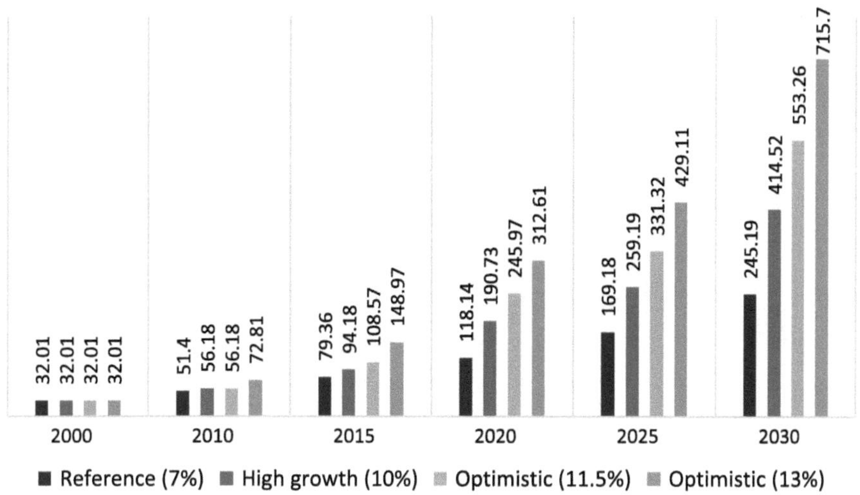

Fig. 2.5 Projections of total energy demand (Mtoe) in Nigeria

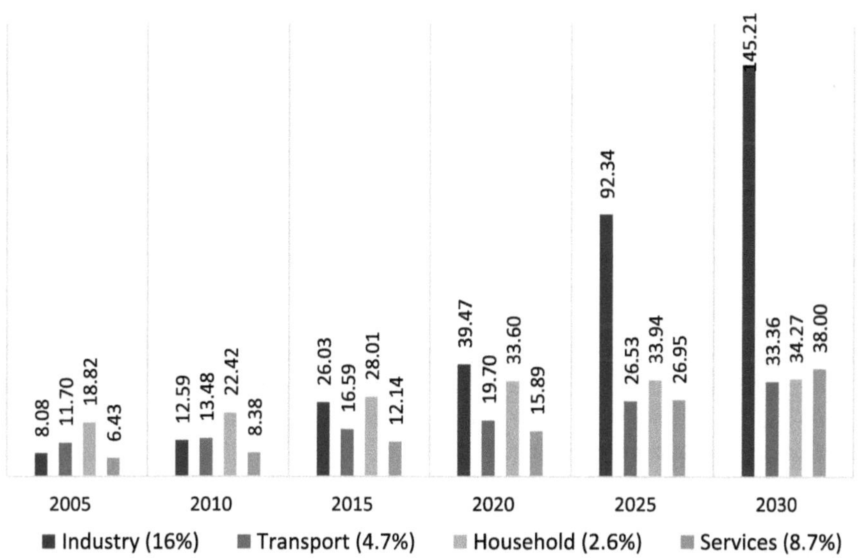

Fig. 2.6 Total energy demand (Mtoe) based on 10 % GDP growth rate

activity. The service sector is projected to have the second highest growth rate of 8.7 %, while the residential sector is expected to have the lowest average growth rate of 2.6 %. These projections for increasing energy demand can only be met if effective policies are put in place.

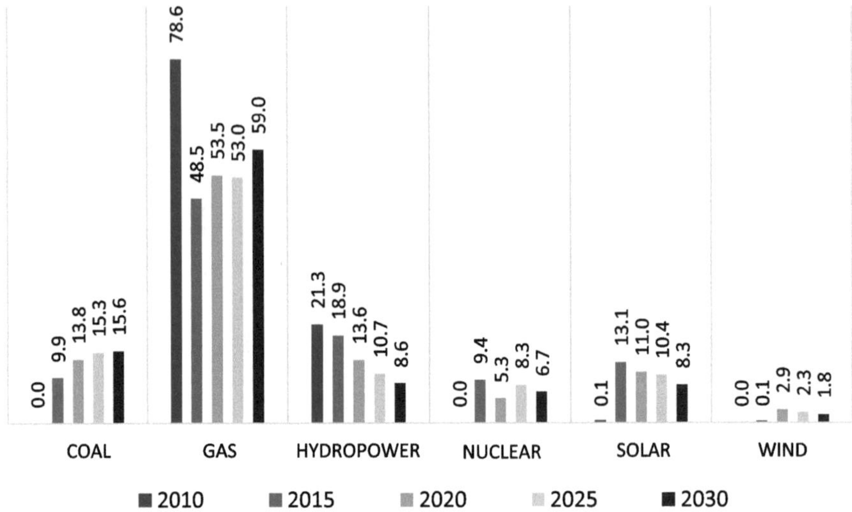

Fig. 2.7 Nigeria's current and future electricity generation capacity (%) by fuel (Reference case)

The ECN undertook a study using MESSAGE[17] under the auspices of the IAEA[18] to ascertain the future fuel mix for the diversification of Nigeria's electricity supply. The study used six different types of fuel for the optimization: coal, natural gas, hydropower, nuclear, solar, and wind energy. Oil was not considered in the optimization because of its use for export to the international energy market, meeting domestic energy demand, and no current plans by the government to establish oil power plants in the future. The results are shown in Fig. 2.7.

These results were based on the reference case scenario, and show that energy from coal and nuclear sources (currently not part of the nation's electricity generation mix) will account for 15.6 and 6.7 % of the total, respectively, by 2030. The proportion of Nigeria's energy generated by hydropower will decrease from 21.3 % in 2010 to 8.6 % by 2030. The high growth and optimistic scenarios follow similar supply patterns (Sambo 2008).

2.2.1.1 Crude Oil

Nigeria produces mostly light sweet crude oils that are predominantly exported to the world market. Table 2.2 shows the characteristic features of the crude oils produced in Nigeria and their ports of sale used for export.

---

[17]Model for the Energy Supply Strategy Alternatives and their General Environmental Impact.

[18]International Atomic Energy Agency.

Table 2.2 Characteristic features of Nigeria's crude oil

Product name	Sulfur content (as % of mass)	API gravity (in degrees)	Ports of sale
Agbami	0.044	47.2	Offshore
Amenam Blend	0.12	38.2	Unity FSO/Odudu Terminal
Amenam/Mars Blend	0.94	33.5	–
Antan Blend	0.27	26.4	Knock Taggart FPSO/Antan Terminal
Bonga	0.26	29.1	Bonga FPSO
Bonny Light	0.16	33.4	Bonny
Brass River	0.16	–	Brass River Terminal
EA Crude	0.08	35.1	Sea Eagle FPSO
Erha	0.12	31.8	Erha FPSO
Escravos	0.17	34.2	Escravos
Forcados (to Europe)	0.16	30.8	Forcados
Odudu	0.15	30.5	–
Okono	0.15	30.50	–
Oso Condensate	0.06	45.7	Qua Iboe
Qua Iboe	0.14	36.3	Qua Iboe
Ukpokiti	0.14	36.3	–
Yoho Crude	0.08	39.3	Falcon FPSO

Source Energy Intelligence (2015)

In Nigeria, commercial production of crude oil began in 1958 based on proven recoverable reserves of 1.48×10^6 billion tons. Production rose from an initial quantity of 3.1 million metric tons to 20.3 million tons in 1960, 54.2 million tons by 1970, and 104.1 million tons in 1980, all in response to demand from international markets rather than from domestic demand. On average, local consumption accounted for just 3 % of production, while the remaining 97 % was exported. Since 1980, three domestic petroleum refineries have supplied petroleum products for local consumption: the Kaduna Refinery with a capacity of 110,000 bbl/d (barrels per day), Port Harcourt Refinery with a capacity of 210,000 bbl/d, and Warri Refinery with a capacity of 125,000 bbl/d (Oyedepo 2014).

As shown in Fig. 2.8, the production of crude oil in Nigeria increased rapidly between 1980 and 2012; however, the rate of increase was dependent on the economic and geopolitical situations in both producing and consuming countries. Nigeria's current production capacity of 2.4 million bbl/d remains low because of problems in the Niger Delta[19] and OPEC production restrictions. However, projections have placed future (2030) production at over 5.0 million bbl/d (Ajao et al. 2009).

[19]The Niger Delta region has been known for militant activity, but this activity was halted by the intervention of the late Nigerian President Umaru Yar'Adua who granted amnesty to the militants.

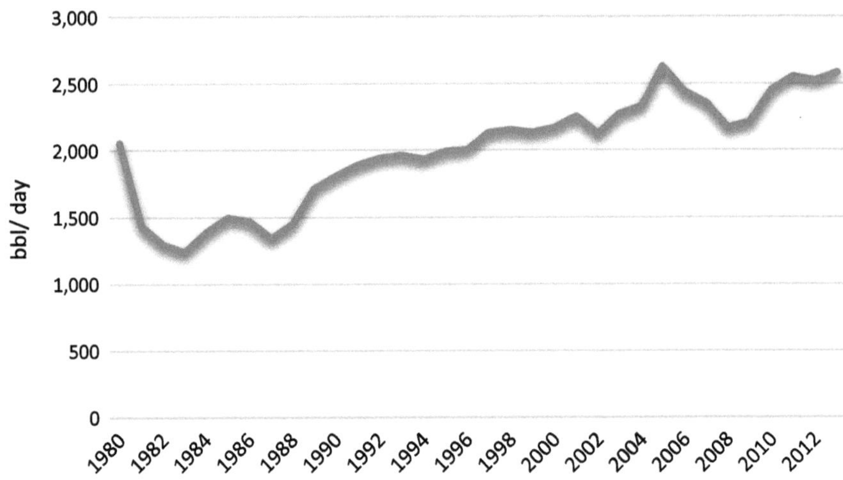

Fig. 2.8 Crude oil production in Nigeria. (EIA 2015a)

Crude oil production reached its peak in 2005, but has subsequently declined significantly because of the activities of militants in the Niger Delta region. These activities came to a halt in 2009 when amnesty was granted to the militants and by 2010, oil production began to increase as oil companies began operating at full capacity. The Nigerian government also took drastic measures to attract investment in deep-water acreage in order to diversify the location of oil fields and increase oil production. This has resulted in the production of an additional 800,000 bbl/d since 2003. However, crude oil production declined from 2011 to 2012 because of heavy floods and supply disruptions.

In addition to the challenges faced by the government, the indigenes of the Niger Delta region suffer from the effects of environmental damages resulting from pipeline vandalism. When pipelines are vandalized, crude oil is stolen to supply illegal refineries. The result is environmental degradation and the risk of a pipeline explosion for local communities. The rates of domestic production and export of crude oil did not improve significantly between 2006 and 2014 (See Fig. 2.9) because of the issue of crude oil theft.

The oil price in Nigeria has been in line with the OPEC price and has fared well over the years, reaching its peak in 2008, but declining in 2009 (Fig. 2.10). The price of oil in Nigeria was US$102.33 per barrel in August 2014; however, with the recent fall in the crude oil price,[20] which is US$53 for West Texas Intermediate (WTI) and US$57.33 for Brent crude, the country's economy will be

[20]According to www.oilprice.net on 2nd January 2015.

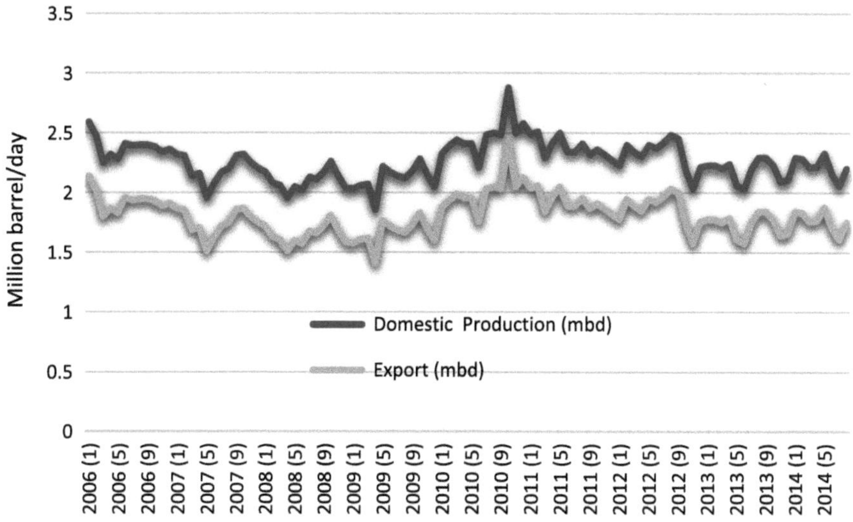

Fig. 2.9 Rates of domestic production and export of crude oil in Nigeria (CBN 2015a)

Fig. 2.10 Crude oil prices in Nigeria. *Source* CBN (2015b)

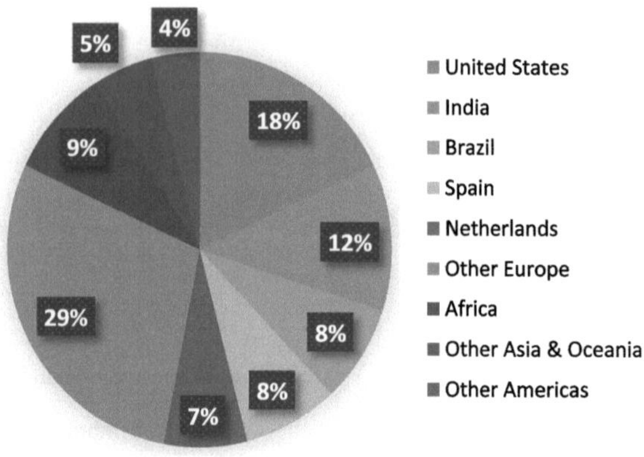

Fig. 2.11 Crude oil and condensate exports in Nigeria in (2012) (EIA 2015b)

adversely affected. The previous rise in the crude oil price was due to high oil consumption in countries such as China and India, in conjunction with conflict in key oil exporting countries such as Libya.[21]

High oil prices induced companies in Canada and the United States (US) to start drilling for new hard to extract crude in North Dakota's shale formations and Alberta's oil sands. This has resulted in a "Price-War" between OPEC and the US (VOX 2015). The US, which was the largest single importer of crude oil from Nigeria (Fig. 2.11) in 2012, ceased importing oil at the end of 2014. Other countries have also reduced their oil imports from Nigeria, including the countries of the European Union.

Fuel subsidies cost the Nigerian government US$8 billion in 2011 alone, which constituted 30 % of federal government expenditure, 4 % of the country's GDP, and 118 % of the capital budget. However, the Nigerian government removed the fuel subsidy on the 1st of January 2012. Subsequently, oil consumption in Nigeria has increased, as shown in Fig. 2.12. Downstream industries in Nigeria, which include domestic refineries and various petrochemical industries, use the crude oil produced in Nigeria. These refineries produce products such as linear alkyl benzene, benzene, heavy alkylate, and deparaffinated kerosene for domestic consumption (NNPC 2015a).

2.2.1.2 Natural Gas

The estimated proven reserves of natural gas in Nigeria stand at 182 trillion cubic feet (TCF) with a mean gauge pressure of about 12 bar, a calorific value of 35 mJ/m^3, and a mean specific volume of 1.56×10^{-3} m^3/kg. In 2012, the production rate was about 1.35 TCF of dry natural gas (Fig. 2.13), making Nigeria the 25th largest producer of dry natural gas in the world (ECN 2007; EIA 2015d, e).

[21] www.useconomy.about.com/od/commoditiesmarketfaq/p/high_oil_prices.htm.

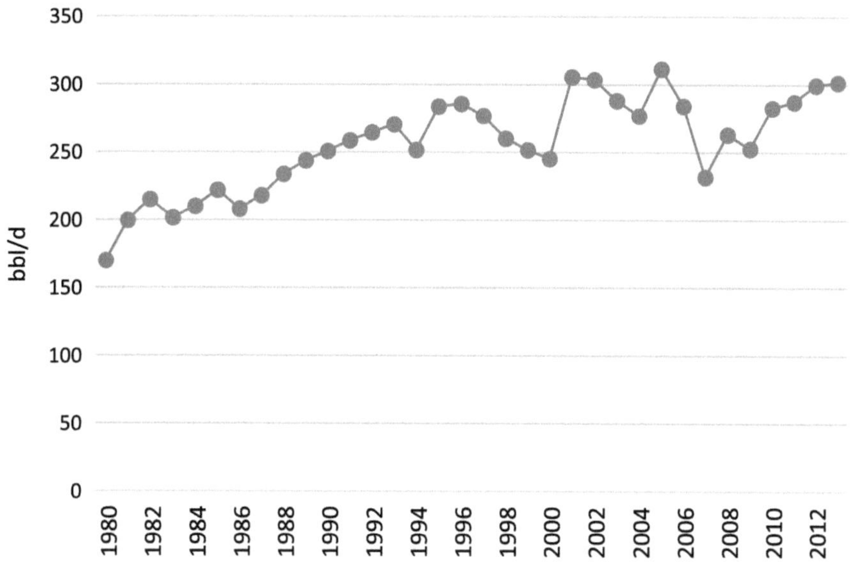

Fig. 2.12 Crude oil consumption in Nigeria (EIA 2015c)

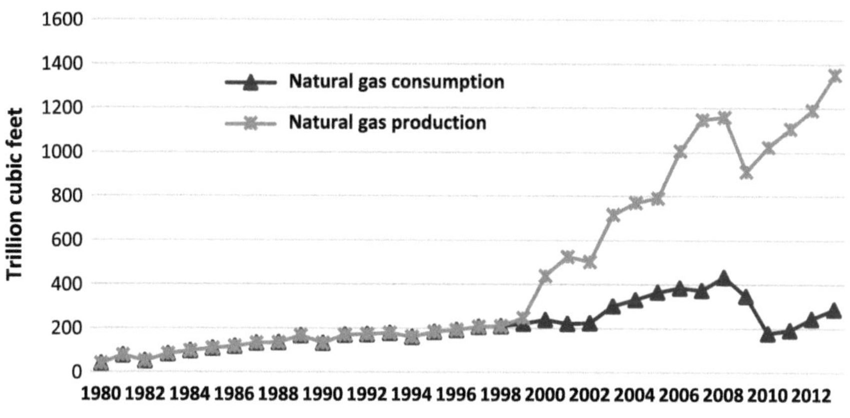

Fig. 2.13 Natural gas production and consumption in Nigeria (CIA 2015a)

Natural gas reserves are located in the Niger Delta region of Nigeria (South-South). In the past, Nigeria flared about 73 % of its gas because of poor infrastructure, which placed Nigeria second in the list of gas-flaring countries (Fig. 2.14).

However, because of the efforts of the Nigerian government to reduce gas flaring through the financing and provision of relevant infrastructure to use the previously flared gas, Nigeria is now 365th on the list (Ibitoye 2014). Infrastructure that uses the previously flared gas includes the power sector, which accounts for 80 % of the total domestic consumption (Table 2.3, Fig. 2.15) and generates 81 % of the total electricity supply in Nigeria.

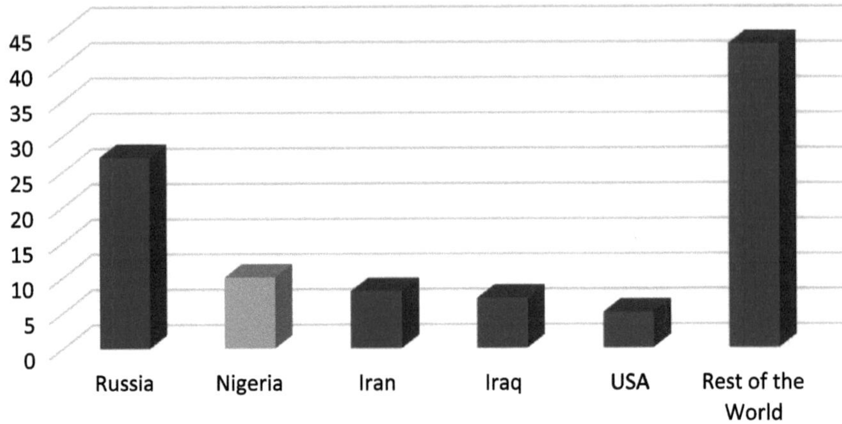

Fig. 2.14 World's top natural gas flaring countries (2012) (EIA 2015f)

Table 2.3 Current and planned power plants in Nigeria and their locations

Power station	Location	Type	Installed capacity (MW)	Year completed
AES Barge	Egbin	SCGT	270	2001
Aba	Aba, Abia State	SCGT	140	2012
Afam IV–V	Afam, Rivers State	SCGT	726	1982
Afam VI	Afam, Rivers State	CCGT	624	2009
Alaoji (NIPP)	Abia State	CCGT	1074	2013
Calabar (NIPP)	Cross River State	SCGT	561	2014
Egbema (NIPP)	Imo State	SCGT	338	2013
Egbin	Egbin	Gas-fired steam turbine	1320	1986
Geregu 1	Geregu, Kogi State	SCGT	414	2007
Geregu 11 (NIPP)	Geregu, Kogi State	SCGT	434	2013
Ibom (IPP)	Ikot Abasi	SCGT	190	2009
Ihorbor (NIPP)	Benin City	SCGT	450	2013
Okpai	Okpai	CCGT	480	2005
Olorunsogo	Olorunsogo	CCGT	336	2007
Olorunsogo 11	Olorunsogo	CCGT	675	2012
Omoku	Omoku	SCGT	150	2005
Omoku 11 (NIPP)	Omoku	SCGT	225	2013
Omotosho 1	Omotosho	SCGT	336	2005
Omotosho 11 (NIPP)	Omotosho	SCGT	450	2012
Sapele	Sapele	Gas-fired steam turbine	1020	1981

(continued)

Table 2.3 (continued)

Power station	Location	Type	Installed capacity (MW)	Year completed
Sapele (NIPP)	Sapele	SCGT	450	2012
Ughelli	Delta State	SCGT	900	1990
Itobe	Kogi State	CFB Technology	1200	2015–2018
Kainji	Niger State	Hydro	800	1968
Jebba	Niger State	Hydro	540	1985
Shiroro	Kaduna State	Hydro	600	1990
Zamfara (Planned)	Zamfara State	Hydro	100	2012
Kano (Planned)	Kano State	Hydro	100	2015
Kiri (Planned)	Benue State	Hydro	35	2016
Mambilla (Planned)	Taraba State	Hydro	3050	2018

Source Emodi and Yusuf (2015a)
NIPP National Integrated Power Project, *SCGT* Single Combined Gas Turbine, *CCGT* Combined Cycle Gas Turbine, *CFB* Circulating Fluidized Bed

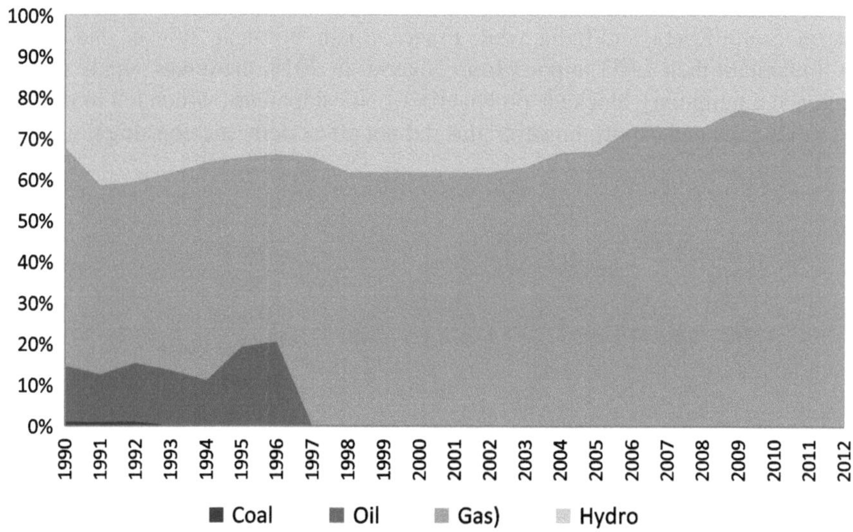

Fig. 2.15 Electricity production by fuel type (%). *Source* Power Holding Company of Nigeria (PHCN) (2010)

Based on the current amount of gas remaining in the natural gas reserves, it is expected that they will last for about 88 years (Kennedy-Darling et al. 2008). In order to improve natural gas activities, in 1988, through the NNPC, the government created a subsidiary and strategic business unit called the Nigerian Gas Company (NGC). This company is responsible for the development of policies for the transmission, distribution, marketing, and pricing of natural gas and all its derivatives to the market within Nigeria and West Africa (NPC 2009).

This implies that a monopoly exists in the Nigerian gas market, which also includes state and private companies such as the electricity companies and multinational companies (known as International Oil Companies; IOCs) including Shell, Chevron, and ExxonMobil who supply natural gas to the NGC (Tallapragada 2009).

Natural gas consumption in Nigeria has been increasing since its discovery and reached its peak in 2008 when a disruption in gas supply occurred (see Fig. 2.13). One of the IOCs (Shell) shut down their plants in order to repair damage to pipelines connected to the Soku plant. This damage was the result of pipeline vandalisation performed by local groups that were siphoning condensate. After five months, the plants re-opened but closed again in 2009 because of operational problems. The Soku plant provides a substantial amount of gas to the NGC's liquefied natural gas (LNG) facility and, because of the supply disruption, a decline in gas production occurred (The Encyclopedia of Earth 2015).

Nigeria's exports of natural gas (in the form of LNG) in 2012 are shown in Fig. 2.16. Europe has reduced its LNG imports from Nigeria since 2012, whereas the US has completely stopped importing from Nigeria because of increasing domestic production. However, imports have increased in Asian countries such as Japan, South Korea, and India, while France, Spain, Portugal, Taiwan, and Turkey still maintain their LNG imports from Nigeria. In 2013, there was supply disruption and a temporary blockade on Nigeria's LNG shipments, which led to a fall in its production and export; however, this did not affect domestic consumption.

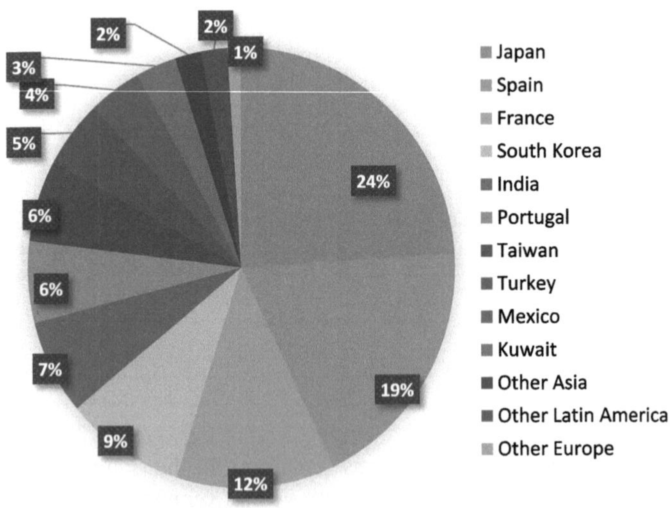

Fig. 2.16 Nigeria's LNG exports (2012) (EIA 2015g)

2.2.1.3 Coal

Coal was the first conventional fuel to be discovered and used in Nigeria. It was discovered in Enugu State, in the south-eastern region of Nigeria in 1909. The first coal mine, the Ogbete drift mine, was opened in 1916 with an output of 24,500 t. Its operation was merged with that of others within the country in 1950 following the formation of a new corporation known as the Nigerian Coal Corporation (NCC). The responsibility of the NCC in holding a monopoly on coal production (including coke mining) and sales was to exploit coal resources. The Polish firm KOPEX was in charge of its management from the NCC's formation until its collapse after the Nigerian Civil War in 1970 (Nwaobi et al. 2005).

Coal production reached a peak of 790,030 tons in 1956 (see Fig. 2.17) and about 70 % of this quantity contributed to Nigeria's energy generation. Production began to decline when oil was discovered in 1956 at Oloibiri in the Niger Delta by Shell-BP, which was the sole concessionaire at that time (NNPC 2015b). The Nigerian Railway Corporation was the largest consumer of coal in the country. The corporation began to convert their railway engines from coal to diesel and gas in 1955. By 1982, production of coal had fallen to 62,830 tons (TOPFORGE 2015). Another set-back in coal production was the conversion of power plants from coal to oil by the National Electric Power Authority (NEPA) (now defunct) and only one coal-fired power plant was left.

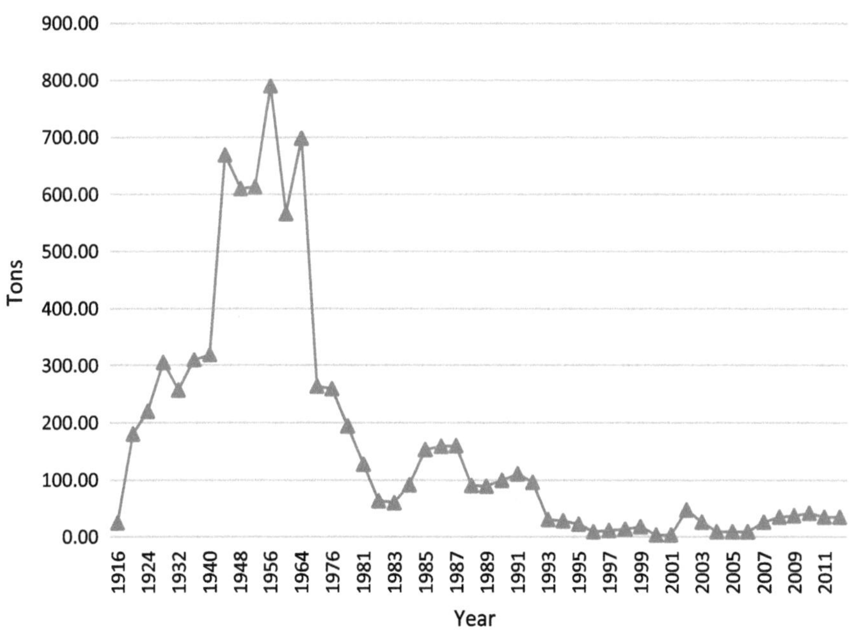

Fig. 2.17 Coal production in Nigeria (Tons). *Source* CIA (2015b)

Following the discovery of coal in commercial quantities (1916–1980), the total cumulative production was about 25.3 million metric tons. By 1980, oil contributed 70 % to electricity generation, compared with 25 % and just 1 % for gas and coal, respectively. In terms of coal consumption, about 95 % of the coal produced in Nigeria was consumed by the Nigerian Railway Corporation, NEPA, and cement companies for heating cements. Coal consumption by the Nigerian Railway Corporation fell to 60 % in 1958, less than 30 % in 1966, and to an insignificant level by 1986. NEPA's coal consumption fell from 30 % in 1966 to an insignificant level in 1970, and the last remaining coal plant was shut down in 1992 (Fig. 2.15) (Enibe and Odukwe 1990).

Because of the loss of its largest consumers, the NCC began to export its desirable low-sulfur-content coal to the United Kingdom and Italy. In 1999, the NCC lost its monopoly over the Nigerian Coal Industry (NCI) following the implementation of the federal government's privatization policy by the then President of Nigeria (i.e. President Olusegun Obasanjo). In 2002, work stopped at all NCC-operated mines and the government established a technical advisory committee to revive the NCI (Odesola et al. 2013a, b).

Current coal reserves in Nigeria are estimated to be 2.75 billion metric tons and the nation's proven reserves stand at 639 million tons. The locations of the coal deposits in Nigeria are mostly in the eastern parts of the country, as shown in Table 2.4. However, the coal reserves have not been fully explored or even marginally developed despite the long history of the coal industry. They are mostly lignite and sub-bituminous, although some are high-volatility bituminous deposits (Akubo et al. 2013).

2.2.2 Non-conventional Energy Resources (Renewable Energy)

Renewable energy plays a vital role in meeting the needs of both rural and urban areas of the country in terms of sustainable development (Hui 1997). The development and proper use of renewable energy should be given high priority, especially now that the issues of climate change and global warming are among the most critical issues discussed by the various governments of the world. Developed and developing countries are now adopting renewable resources in order to achieve energy sustainability (Oyedepo 2012).

Nigeria is blessed with an abundance of renewable energy resources that must be fully harnessed, developed, and properly used. However, the development of renewable energy has so far been slow, and the desperate situation of the energy sector in Nigeria can only be resolved if adequate policies are implemented to attract investors in renewable energy to Nigeria. Nigeria's renewable energy resources are presented in Table 2.5 and discussed in detail below.

Table 2.4 Coal reserves in Nigeria

Mine location	State	Coal type	Estimated reserves (million tonn)	Proven reserves (million ton)	Borehole record	Coal outcrop and seam thickness (m)	Coal depth (m)	Mining approach	Current status
Okpara mine	Enugu	Sub-Bituminous	100	24	20	Many	180	Under-ground	Functional
Onyeama mine	Enugu	Sub-Bituminous	150	40	Many	Many	180	Under-ground	Functional
Ezimo	Enugu	Sub-Bituminous	156	56	4	10 (0.6–2.0)	30–45	Under-ground	
Inyi	Enugu	Sub-Bituminous	50	20	4	(0.9–2.0)	25–48	Open cast, Under-ground	
Amansiodo	Enugu	Bituminous (cokeable)	1000	NA	3	NA	563	Under-ground	
Ogugu/Awgu	Enugu	Sub-Bituminous	NA	NA	Nil	NA	NA	Under-ground	
Okaba	Kogi	Sub-Bituminous	250	73	Many	(0.8–2.3)	20–100	Open cast, Under-ground	Functional
Ogboyoga	Kogi	Sub-Bituminous	427	107	31	17 (0.8–2.3)	20–100	Open cast, Under-ground	
Ogwashi-Uku/Azagba/Obomkpa	Delta	Lignite	250	63	7	4 (3.5)	15–100	Open cast, Under-ground	
Oba/Nnew	Anambra	Lignite	30	NA	2	14 (0.3–4.5)	18–38	Underground	
Ihioma	Imo	Lignite	40	NA	Nil	Many	20–80	Open cast	
Lafia/Obi	Nasarawa	Bituminous (cokeable)	156	21.42	123	–1.3	80	Under-ground	
Owukpa	Benue	Sub-Bituminous	75	57	Many	(0.8–2.3)	20–100	Open cast, Under-ground	Functional
Afikpo/Okigwe	Abia	Sub-Bituminous	50	NA	Nil	NA	20–100	Under-ground	
Afuze	Edo	Sub-Bituminous	NA	NA	Nil	NA	NA	Under-ground	
Ute	Ondo	Sub-Bituminous	NA	NA	Nil	NA	NA	Under-ground	
Lamja	Adamawa	Sub-Bituminous	NA	NA	Nil	NA	NA	Under-ground	
Gandi-Akwati	Plateau	Sub-Bituminous	NA	NA	Nil	NA	NA	Under-ground	
Jamata-Koji	Kwara	Sub-Bituminous	NA	NA	Nil	NA	NA	Under-ground	
Doho	Gombe	Sub-Bituminous	NA	NA	Nil	NA	NA	Under-ground	
Kurumu Pindiae	Gombe	Sub-Bituminous	NA	NA	Nil	NA	NA	Under-ground	
Garin Maiganga	Gombe	Sub-Bituminous	NA	NA	Nil	NA	NA	Under-ground	

Source Essien and Igweonu (2014)

Table 2.5 Renewable energy resources in Nigeria and their potential

Resource type	Reserves		Production	Domestic utilization (natural units)
	Natural units	Energy units (Btoe)		
Small Hydro-power	3500 MW	0.34 (over 40 years)	30 MW	30 MW
Large Hydro-power	11,250 MW	0.8 (over 40 years)	1938 MW	1938 MW
Wind	2–4 m/s at 10 m height (main land)	0.0003 (4 m/s @ 12 % probability, 70 m height, 20 m rotor, 0.1 % land area, 40 years	–	–
Solar Radiation	3.5–7.0 kWh/m^2/day (4.2 million MWh/day using 0.1 % land area)	5.2 (40 years and 0.1 % land area)	6 MWh/day	6 MWh/day
Biomass Fuel wood	11 million hectares of Forest and wood land Excess of 1.2 m ton/day	–	0.120 million ton/day	0.120 million ton/day
Animal waste	211 million assorted animals	–	0.781 million ton of waste/day	None
Energy crops and agricultural residue	28.2 million hectares of arable land (= 30 % of total land)	–	0.256 million ton of assorted crops/day	None

Source ECN (2009)

2.2.2.1 Biomass Energy

Biomass energy refers to energy that is developed from organic materials like scrap lumber, forest debris, crops, manure, and some types of waste residue. Biomass is an indirect form of solar energy because it arises from the process of photosynthesis. Biomass resources found in Nigeria include wood, shrubs, forage grasses, and waste from animals, forestry, agriculture, industry, and municipal areas. Nigeria's biomass resources have been estimated at 88×10^2 MJ.[22] Biomass energy from plants could be used as fuel for small-scale industries or fermented by anaerobic bacteria to produce cheap and versatile biogas (Garba and Bashir 2002).

Fuel wood is the most common form of biomass in Nigeria, with about 80 million m^3 used annually for cooking and various other domestic purposes (Sambo 2005). The energy content of this fuel wood is 6.0×10^9 MJ, out of which only 5 and 12 % are used for cooking and other domestic uses, respectively (Lawal 2007a, b). In addition, increasing demand for wood by the furniture and construction industries is causing a rapid depletion of the biomass resources in Nigeria.

Shrubs and forage grasses have been estimated to produce 200 million tons of dry biomass that could release up to 2.28×10^6 MJ of energy (Vincent and Yusuf 2014). Because of the high dependence on fuel woods for cooking and heating by rural dwellers in Nigeria, 350,000 ha of forest and vegetation are lost annually, although this is much lower than the afforestation rate of 50,000 ha per annum (Sambo 2009a).

However, soil erosion and desert encroachment will result from these activities if the situation is not properly controlled. This could be achieved by discouraging the use of firewood through the introduction of affordable solar stoves. The introduction of a three-stone stove with an efficiency as low as 15 %, which was developed locally by the ECN through its energy research centers at the University of Nigeria in Nsukka and Usman Dan Fodio University in Sokoto, will ensure the reduction of fuel wood consumption (Sambo 2009a).

Biomass is an important renewable energy source, but the sustainability of its production needs to be clearly understood. Nigeria should use its wood, municipal waste, oil palm product, sugar cane, and rice husk resources sustainably for biogas energy production. As has been practiced in South Africa and Malaysia, sugar mill companies in Nigeria could make use of their cane residues and waste, while paper and packaging mills could use their waste biomass to generate process steam (Shaaban and Petinrin 2014). Table 2.6 presents the estimated quantities and energy values of Nigeria's biomass resources.

[22] Nadabo (2010).

Table 2.6 Nigeria's biomass resources, estimated quantities, and energy values

Resources	Quantity (million ton)	Energy value (000 MJ)
Fuelwood	39.1	531.0
Agro-waste	11.244	147.7
Saw dust	1.8	31.433
Municipal solid waste	4.075	–

Source Sambo (2009a)

Biogas

Biogas is produced from the anaerobic digestion of agricultural and animal waste in the absence of air. It has an estimated combustion temperature in the range of 65–750 °C and it is 20 % lighter than air. Biogas is similar to LPG gas because it is colorless or odorless and it burns with a brilliant blue flame. Its caloric value has been estimated to be about 20 MJ/m^3 and it burns with an efficiency of about 55 % in a conventional biogas stove. The gas contains a mixture of carbon IV oxide, hydrogen sulfide, methane, nitrogen, and water vapor (Opeh and Okezie 2011). The raw materials for biogas production include animal dung and waste from industry, farmland, and households.

Biogas constitutes a form of energy suitable for households and the agricultural and industrial sectors of the economy. It is a useful substitute for diesel, fuel wood, charcoal, and kerosene; it reduces GHG emissions, and it has no health risks because it burns clean (Akinbami et al. 2001). In rural areas of Nigeria, suitable feed stock has been identified and is considered economically viable for the production of biogas, including cassava leaves, dung, solid waste, water hyacinth, water lettuce, agricultural residues, urban refuse, and sewage (Akinbami et al. 1996).

Studies have shown that Nigeria produces about 227,500 t of fresh animal waste daily and 20 kg of municipal solid waste per capita annually (Mshandete and Parawira 2009; Oyedepo 2012). About 0.03 m^3 of gas can be produced from 1 kg of fresh animal waste; therefore, 6.8 million m^3 of biogas could be produced daily in Nigeria. Research conducted by Adeoti (1998) showed that a 6.0-m^3 family-sized biogas digester could produce about 2.7 m^3 of biogas per day, which would be sufficient to satisfy the cooking needs of a family of nine persons.

The initial cost of such project was estimated at US$500 (i.e., NGN 80,100 in Nigeria Naira); annual expenditure was estimated as NGN 11,200, while the benefit was estimated as NGN 25,000. Although the project appears to have good economic potential, it might be too expensive for the low-income earners who reside mostly in rural areas. If measures are not taken to lower the costs or assist the low-income earners economically, low-income households might not accept the use of biogas (Garba and Bashir 2002).

It is of great importance for Nigeria's government to establish some biogas plants to help the development of the country's energy sector, because the technology can generate energy rapidly as the raw materials needed to feed the biogas plants are relatively abundant across the country (Opeh and Okezie 2011).

In addition to the use of biogas for household consumption and electricity generation, other areas such as the transport sector could benefit from this renewable option. The production of biogas in Nigeria would not only develop the energy sector but also aid in the reduction of urban waste.

2.2.2.2 Hydro Energy

Hydro energy technology is dependent on the potential energy difference between the levels of water in reservoirs, dams, or lakes and their discharge tail water levels downstream. A water turbine, which converts the potential energy of the water to shaft rotation, is coupled to a suitable generator to produce electricity (Sambo 2005).

In Nigeria, hydro energy technology is currently the prominent commercial renewable energy technology in the country's electricity supply mix. Economy of scale has enabled the development of large-scale hydropower technology to account for a large proportion of the total commercial renewable energy resources for electricity generation under greenhouse gas (GHG) emission constraints (Balogun 2010).

Apart from the problem of relative water levels, hydropower can supply uninterrupted electric power. Nigeria's total hydropower potential stands at 14,750 MW, but only 1930 MW (i.e., 14 %) is currently generated at Kanji, Shiroro, and Jebba, which represents about 30 % of the gross installed grid-connected generation capacity in Nigeria (CBN 2005). This assessment is based only on large-scale hydropower that was in operation before the 1973 oil crisis.

Clearly, Nigeria's hydropower potential has not been fully exploited. However, small hydropower (SHP) has recently received considerable global attention. This attention is because of the inherent advantages of SHP in reducing environmental impact, minimizing civil works, and offering the possibility of combining power generation with flood prevention, irrigation, and the development of fisheries. Nigeria's current SHP generation is estimated at 3500 MW, which represents about 23 % of the entire national hydro potential, as shown in Table 2.5.

A study undertaken in 12 states and 4 river basins revealed over 278 unexploited SHP sites with a total potential of 734.2 MW (Aliyu and Elegba 1990). Three of the states surveyed, Kano, Sokoto, and Plateau, had installed operating SHP generators with a total capacity of 30 MW. The Nigerian Electricity Supply Company (NESCO) is currently generating 21 MW from six other sites in Plateau state. Currently, about 5 % of the available SHP capacity is exploited, while other SHP sites have been set aside for future development. However, out of the total potential of 734.2 MW, only 32 MW have been developed. Table 2.7 shows the SHP potential in Nigeria, Table 2.8 presents the existing SHP schemes (Sambo 2009b), and Fig. 2.18 shows the various water ways within the country.

Table 2.7 Small hydropower potentials in Nigeria

State (Pre 1980)	River basin	Total sites	Hydropower potential		
			Developed (MW)	Undeveloped (MW)	Total capacity (MW)
Sokoto	Sokoto-Rima	22	8.0	22.6	30.6
Katsina	Sokoto-Rima	11		8.0	8.0
Niger	Niger	30		117.6	117.6
Kaduna	Niger	19		59.2	59.2
Kwara	Niger	12		38.8	38.8
Kano	Hadeija–Jamaare	28	6.0	40.2	46.2
Borno	Chad	28		20.8	20.8
Bauchi	Upper Benue	20		42.6	42.6
Gongola	Upper Benue	38		162.7	162.7
Plateau	Lower Benue	32	18.0	92.4	110.4
Benue	Lower Benue	19		69.2	69.2
Cross Rivers	Cross Rivers	18		28.1	28.1
Total		277	32	702.2	734.2

Source Sambo (2009b)

Table 2.8 Small hydropower schemes in existence in Nigeria

River	State	Installed capacity (MW)
Bagel I	Plateau	1
Bagel II	Plateau	2
Ouree	Plateau	2
Kuna	Plateau	8
Lere	Plateau	4
Lere	Plateau	4
Bakalori	Sokoto	3
Tiga	Plateau	6
Total		30

Source Sambo (2009b)

2.2.2.3 Solar Energy

Solar energy is the most promising renewable energy source because of its apparent limitless potential. The sun radiates energy at the rate of about 3.8×10^{23} kW/s. Most of this energy is transmitted radially as electromagnetic radiation, reaching the boundary of Earth's atmosphere at about 1.5 kW/m^2. After traversing the atmosphere, a square meter of Earth's surface can receive as much as 1 kW of solar power, or about 0.5 kW on average during daylight hours. This huge energy resource is available for about 26 % of the day (Muhammad 2012).

Solar energy can provide cheap and abundant energy for communities whose connection to the utility grid might not be economical because they are located

Fig. 2.18 Nigerian water ways. *Source* Safty4sea (2014)

too far from the nearest grid-connection point. Solar energy is therefore a very good alternative source of energy in the rural areas of Nigeria. It could aid the rapid development of small-scale industries and reduce rural–urban migration (Ojosu 1990).

Nigeria is located within a high sunshine belt and solar radiation is well distributed. The annual average total solar radiation varies from about 25.2 MJ/m^2/day (7.0 kWh/m^2/day) in northern regions to about 12.6 MJ/m^2/day (3.5 kWh/m^2/day) in southern parts. Assuming an average of 18.9 MJ/m^2/day (5.3 kWh/m^2/day), Nigeria has an estimated 17,459,215.2 million MJ/day (17.439 TJ/day) of solar energy arriving over its 923,768 km^2 land area. The annual average intensity is 6898.5 or 1934.5 kWh/m^2/year (Vincent and Yusuf 2014).

As the average sunshine per day is 6.5 h, the annual solar energy available is about 27 times that of the country's total fossil fuel resource, and it is over 115,000 times the electrical power generated (Augustine and Nnabuchi 2009). This implies that about 3.7 % of the land area in Nigeria could collect an amount of solar energy equivalent to the conventional energy reserves within the country (Shaaban and Petinrin 2014). This is in agreement with a study by Oji et al. (2012), considering that the minimum harnessing of solar power in some parts of Nigeria would be more than enough to power an average three-bedroom flat and two-room apartment that use low-power appliances. The monthly averages of daily solar radiation for 28 states in Nigeria over 25 years are presented in Table 2.9, and the annual average daily sun in Nigeria is illustrated in Fig. 2.19.

Table 2.9 Maximum, minimum, and yearly average global solar radiation ($kWh/m^2/day$)

Stations	Location lat. 1 N	Location long 1E	Altitude (m)	Max[a]	Min[b]	Monthly average
Abeokuta	7.25	3.42	150	4.819	3.474	4.258
Abuja	9.27	7.03	305	5.899	4.359	5.337
Akure	7.25	5.08	295	5.172	3.811	4.485
Azare	11.8	10.3	380	6.028	5.022	5.571
Bauchi	10.37	9.8	666.5	6.134	4.886	5.714
Benin City	6.32	5.6	77.52	4.615	3.616	4.202
Calabar	4.97	8.35	6.314	4.545	3.324	3.925
Enugu	6.47	7.55	141.5	5.085	3.974	4.539
Ibadan	7.43	3.9	227.23	5.185	3.622	4.616
Ilorin	8.48	4.58	307.3	5.544	4.096	4.979
Jos	9.87	4.97	1285.58	6.536	4.539	5.653
Kaduna	10.6	7.45	645.38	6.107	4.446	5.672
Kano	12.05	8.53	472.14	6.391	5.563	6.003
Katsina	13.02	7.68	517.2	5.855	3.656	4.766
Lagos	6.58	3.33	39.35	5.013	3.771	4.256
Lokoja	7.78	6.74	151.4	5.639	4.68	5.035
Maiduguri	11.85	13.08	383.8	6.754	5.426	6.176
Makurdi	7.73	8.53	112.85	5.656	4.41	5.077
Minna	9.62	6.53	258.64	5.897	4.41	5.427
New Bussa	9.7	4.48	152	5.533	4.15	4.952
Nguru	12.9	10.47	342	8.004	6.326	6.966
Obudu	6.63	9.08	305	5.151	3.375	4.224
Oweri	5.48	7.03	120	4.649	3.684	4.146
Port Harcourt	4.85	7.02	19.55	4.576	3.543	4.023
Serti	7.5	11.3	610	4.727	3.972	4.488
Sokoto	13.02	5.25	350.75	6.29	5.221	5.92
Wari	5.52	5.73	6.1	4.237	3.261	3.748
Yola	9.23	12.47	186.05	6.371	4.974	5.774

Source Okoro et al. (2007)
[a]Average for the months of March, April and May
[b]Average for the months of July and August

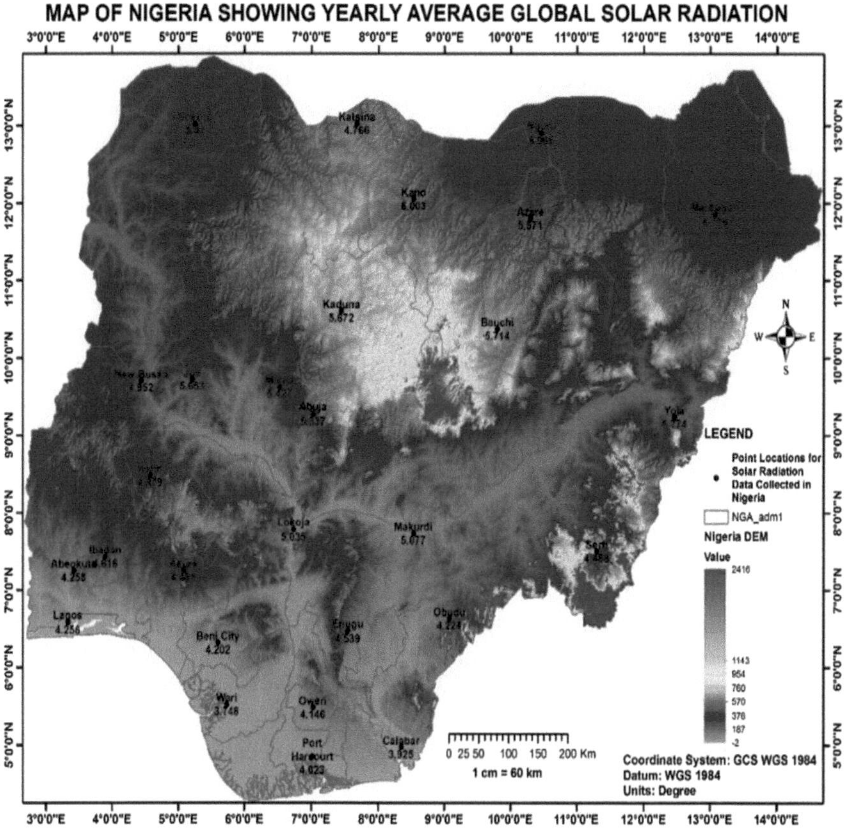

Fig. 2.19 Yearly average of solar radiation in Nigeria. *Source* Nwoko (2015b)

The levels of solar energy awareness and acceptance have already gained ground in northern parts of Nigeria, as presented in a survey performed by Shehu (2012). Other studies, surveys, and pilot projects have been undertaken by the Sokoto Energy Research Centre and the National Centre for Energy Research and Development under the supervision of the ECN. They have implemented solar PV water pumping and electrification, and solar thermal installations such as solar cooking stoves, crop drying facilities, incubators, and chick-brooding systems. However, solar technology has not penetrated into the deep rural areas, especially the off-grid areas, where candles and kerosene lamps are still being used for lighting homes at night. An effective policy should be created to foster the development of solar energy across Nigeria to help reduce poverty in rural areas of the country.

2.2.2.4 Wind Energy

Wind is a natural phenomenon related to the movement of air masses caused primarily by differential solar heating of the earth's surface. The seasonal variation in the energy received from the sun affects the strength and direction of the wind. The ease with which aero-turbines transform the energy of moving air into rotary mechanical energy lends itself to the conversion of wind energy to electricity. For many years, wind energy has been used for pumping water and milling grain (Sambo 1981).

Wind energy generation has gained worldwide recognition and it is the fastest growing renewable energy market in the world. The global cumulative installed capacity of wind power has increased steadily from 6100 MW in 1996 to 158,505 MW in 2009, and was expected to be over 238 GW by the end of 2014, a target that will aid the reduction of GHG emissions (WWE 2008). Currently, 82 countries generate electricity from wind energy, 49 of which have increased their installed capacity since 2009.

In 2009, Egypt, Morocco, and Tunisia were the leading countries in Africa for wind energy with installed capacities of 430, 253, and 54 MW, respectively (Adaramola and Oyewola 2011). In Nigeria, wind measurements at 10-m height show that some sites have wind speeds between 1.0 and 5.1 m/s. These wind speeds can be classified into four regimes: >4.0 m/s, 3.1–4.0 m/s, 2.1–3.0 m/s, and 1.0–2.0 m/s. Therefore, Nigeria is located within a moderate wind[23] regime. The wind speed in southern Nigeria is relatively low, except for coastal regions and offshore, where the high wind speeds indicate great potential for exploiting wind energy (Vincent and Yusuf 2014).

A study undertaken by the ECN revealed that the total exploitable wind energy reserves at 10-m height might vary from 8 MWh/year in Yola to 51 MWh/year in the mountainous area of Jos, and possibly even as high as 97 MWh/year in Sokoto (Sambo 1981). In addition to the study performed by the ECN, many indigenous researchers (Adekoya and Adewale 1992; Fagbenle and Karayiannis 1994; Ngala et al. 2007; Mnse and Ojo 2009; Felix et al. 2012) have analyzed wind data from various parts of the country and these data include wind speeds and power flux densities. Furthermore, the wind energy potential and the conditions that must be met before a wind turbine can be connected to the utility grid have also been studied in the literature (Adekoya and Adewale 1992; Felix et al. 2012; Shaaban and Petinrin 2014).

Adekoya and Adewale (1992) presented estimates of potential wind speeds at 10 selected sites within the country. The results were compared with wind

[23]Moderate wind or breeze according to the Beaufort wind force scale gives the sea condition as small waves with breaking crests, fairly frequent whitecaps and land condition as dust, loose paper raised, small trees branches begin to move (See National Meteorological Library and Archive fact Sheet 6-The Beaufort Scale-PDF).

Table 2.10 Summary of the measured data of annual wind speeds

Site	Land-use type	Altitude (m a.s.l)	Height (m)	Wind speed (m/s)		Differences (%)
				Measured	KLIMM	
Enugu	Complex landscape	466	30	4.6	4.4	–4.3
Jos	Complex landscape	1344	30	5.2	5.1	–1.9
Pankshin	Complex landscape	1355	40	4.9	4.7	–4.1
Sokoto	Plain surface	352	30	5.4	5.2	–3.7
Kano	Plain surface	340	30	4.9	5.1	4.1
Gumel	Plain surface	393	30	4.1	4.2	2.4
Maiduguri	Plain surface	373	30	4.7	4.6	–3
Ibi	River valley	300	30	3.6	3.3	–8.3
Gembu	Highly complex landscape	1800	40	5	5.2	1
Lagos	Coastal area	2	30	4.7	4.9	4.3

Source Adekoya and Adewale (1992), Felix et al. (2012)

Table 2.11 Estimated gross energy yield

Site	Gross energy yield measurement (MWh)		
	Model FL 100, 100/20 Rotor dia. 21.0 m Hub height 34.5 m 100/20	Model FL250, 250/50 Rotor dia. 29.5 m Hub height 42.0 m 250/50	Model V52, 850/52 Rotor dia. 52.0 m Hub height 44.0 m
Enugu	92.9	217.9	734.20
Jos	129.6	299	1025.80
Pankshin	117.1	272.1	936.60
Sokoto	153.5	358.8	1235.80
Kano	116.3	281.2	963.60
Gumel	73.4	197.2	681.40
Maiduguri	102.7	262.2	906.10
Ibi	49.8	141.3	481.20
Gembu	112.9	253.9	855.30
Lagos	129.3	386.1	1402.80

Source Adekoya and Adewale (1992), Felix et al. (2012)

speeds calculated by the Mainz climate model, which revealed a discrepancy of –4.3 to 4.1 %, which is within acceptable error limits. The results, presented in Tables 2.10 and 2.11 list the estimated gross energy yield, while Table 2.12 presents the estimated wind energy potential of some selected states in Nigeria. Figure 2.20 shows the potential of wind energy locations in Nigeria.

According to Shaaban and Petinrin (2014), assuming a medium generation capacity of 5 MWh/km^2 with (a) a 30 % capacity factor and (b) using only 1 % of

Table 2.12 Estimated wind energy potentials

Selected state	Area (km²)	Wind area (%)	Effective wind area (km²)	1 % area (km²)	Potential capacity (MW)	Potential generation (MWh/year)
Adamawa	37,957	45	170,80	171	854	2244
Bauchi	48,197	50	24,098	241	1204	3166
Borno	72,767	100	72,767	728	3638	9561
Gombe	17,428	100	17,428	174	871	2290
Jigawa	23,415	100	23,415	234	1170	3076
Kaduna	44,217	60	26,530	265	1326	3486
Kano	20,389	90	18,350	184	917	2411
Katsina	23,822	100	23,822	238	1191	3130
Kebbi	36,320	25	9080	91	454	1193
Plateau	26,539	90	23,885	239	1194	3138
Sokoto	32,146	90	28,931	289	1446	3801
Taraba	58,180	40	23,672	237	1183	3110
Yobe	44,880	100	44,880	449	2244	5897
Zamfara	33,667	80	26,933	269	1346	3539
Total				3809	19,043	50,046

Source Adckoya and Adcwalc (1992), Fclix ct al. (2012)

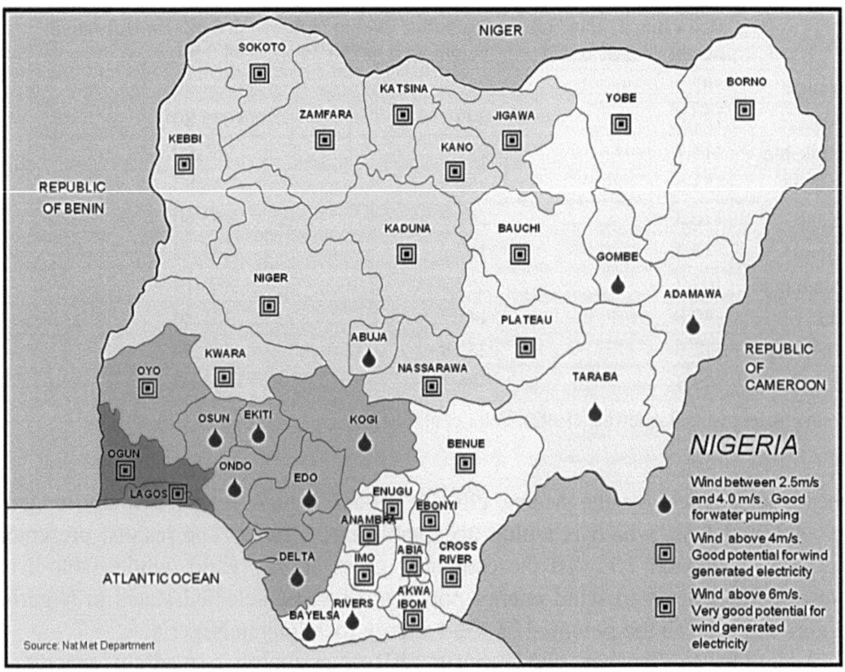

Fig. 2.20 Wind energy potentials and locations in Nigeria. *Source* Nee Nigeria (2015)

Table 2.13 Wind energy estimates at 25 m height

Site	Mean wind speed at 25 m level (m/s)	Monthly mean wind energy (KWh)	Annual wind energy (KWh)	Annual wind energy from a wind turbine (KWh)	
				10 m blade diameter	25 m blade diameter
Benin City	2.135	2.32	27.86	2,18.81	13,673.78
Calabar	1.702	1.12	13.42	1.053.69	6587.53
Enugu	3.372	7.83	93.91	7375.75	46,097.96
Ibadan	2.62	4.15	49.78	3909.70	24,436.19
Ilorin	2.078	1.23	14.73	1157.06	7230.57
Jos	4.43	16.05	192.64	15,129.60	94,559.98
Kaduna	3.605	9.91	188.88	936.81	58,355.08
Kano	3.516	8.57	102.86	8078.61	50,491.28
Lagos (Ikeja)	2.671	4.36	52.32	4099.78	25,682.52
Lokoja	2.235	2.6	31.21	4451.23	15,320.17
Maiduguri	3.486	8.42	101.01	7933.61	49,583.17
Mina	1.589	1.05	12.60	989.60	6185.01
Makurdi	2.689	4.44	53.27	4183.51	26,148.85
Nguru	4.259	14.48	173.74	14,645.19	85,284.42
Oshogbo	1.625	1.07	12.81	1006.60	6288.09
Port Harcourt	2.64	4.17	49.98	3925.48	24,533.88
Potiskum	3.636	9.44	113.25	8894.35	55,591.46
Sokoto	4.476	16.47	197.68	15,525.75	97,035.94
Warri	2.027	2.02	24.20	19,00.66	11,879.15
Yelwa	3.36	7.76	93.13	7314.88	45,714.59
Yola	1.824	1.45	17.34	1.361.88	8511.75
Zaria	2.891	5.32	63.88	5,017.26	31,357.02
Total		134.23	1680.50	120,078.90	790,548.39

Source Adekoya and Adewale (1992), Felix et al. (2012)

the effective wind area of the selected states, Nigeria has the potential to generate about 50,046 MWh/year of electricity. The detailed potentials and wind energy densities at 25-m height for 22 selected states in Nigeria are presented in Table 2.13.

Currently, no commercial wind power plants are connected to the national grid in Nigeria. The few power plants that exist are those installed in the 1960s in five northern states and the 5-kW wind electricity conversion system installed in Sayyan Gidan Gada in Sokoto state. However, the latest development in wind energy generation in Nigeria is the ongoing installation of wind turbines in Katsina state, which is expected to generate 20 MW of electricity on completion. The progress of harnessing the potential of wind energy is too slow. Most communities in northern parts of the country are not connected to the electricity grid and therefore, the federal government must do more regarding the exploitation of wind power, especially in the northern areas where the wind speed is high.

2.3 Primary Energy Supply in Nigeria

According to the International Energy Agency, Nigeria's total primary energy supply for 2012 was 126,097 ktoe (excluding electricity trade and oil product import), and this was almost double the amount in 1990 (i.e. 69810 ktoe). In the primary energy supply mix, biofuels and waste had a percentage share of 85.8 %, while crude oil, natural gas, and hydro had 3.8, 10 and 0.4 % respectively. Crude oil exported out of the country in the same year (2012) was 126,413 ktoe, natural gas was 21,032 ktoe, while oil products were 755 ktoe. Oil products of about 8440 ktoe were also imported, so as to augment supply and meet the growing demand for petroleum products.

As previously stated, Nigeria is among the world's largest oil and gas producers, but depends heavily on the importation of oil products due to the low production capacity of the nation's crude oil refineries. However, with the abundant energy resources available in Nigeria, the largest energy consumption has been biofuels and waste, of which fuel wood or firewood is the primary biofuel source. This is shown in Fig. 2.21, where biofuels and waste have been the main fuel for energy supply from 1990 to 2012. The primary energy supply has been on the decrease, while natural gas has gradually been increasing. This may be due to the price of oil which was higher than natural gas, hence the increase in gas supply. Also, with the increasing construction of gas power plants in Nigeria, the supply of natural gas is expected to increase more than the 2012 values.

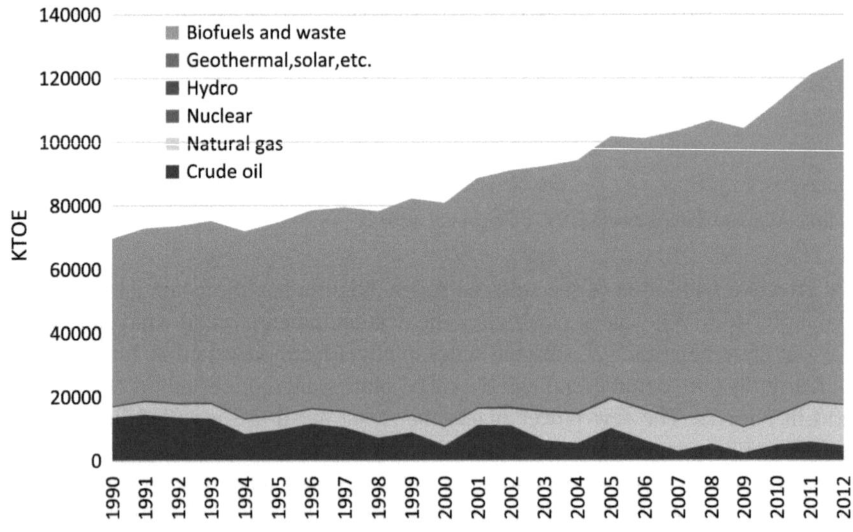

Fig. 2.21 Nigeria's total primary energy supply (1990–2012). *Source* IEA (2015a)

2.4 Primary Energy Consumption

Among the primary energy consumption in Nigeria, biofuels and waste comprise about 89 % of the total primary energy consumption in 2012 alone, while oil products which are mainly imported and natural gas comprise 9 and 2 % respectively (Fig. 2.22). Continuous utilization of biofuels (especially firewood) has been the single most important factor for the increase in deforestation and increase in desert encroachment in most parts of Nigeria. The rate of forest tree replacement is lower than fuel wood (from forest trees) consumption in Nigeria. This will be on the increase as the rate of population in the rural part of Nigeria tends to increase; the total population of Nigeria is forecasted to increase to about 400 million by 2050 (see Fig. 2.4).

According to the study by Zaku et al. (2013), the increase in firewood consumption can be attributed to the increase in poverty, as firewood is usually free, sometimes affordable, and easily accessible to the rural dwellers. However, some studies such as those by Naibbi and Healey (2013), Mandelli et al. (2014), Eludoyin (2015) and Oladimeji et al. (2015) show that firewood consumption is higher in the northern part of Nigeria, while the southern part diversifies energy consumption.

Other studies such as Audu (2013) who used descriptive and comparative methods to analyze the fuel consumption in Nigeria and desertification, shows that in Nigeria, firewood is about the only source of domestic fire in the desert-prone states leading to desertification. This is also reported by Al-Amin (2014) study as it extends the issue to other parts of the country and suggested an immediate action by the Nigerian Government. The total primary energy consumption in Nigeria by sector is shown in Fig. 2.23.

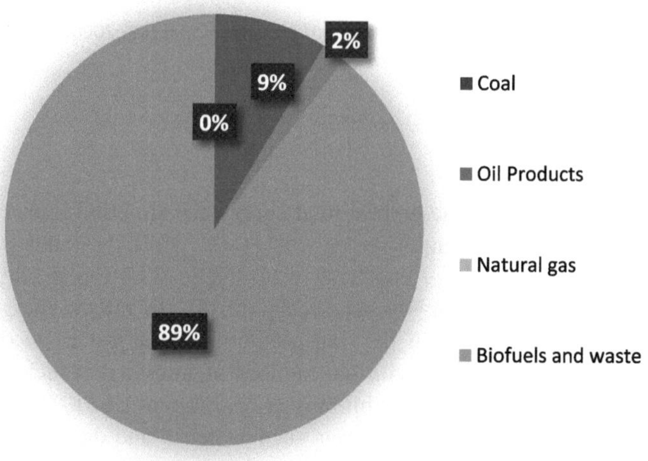

Fig. 2.22 Primary energy consumption in Nigeria by fuel (2012). *Source* IEA (2015b)

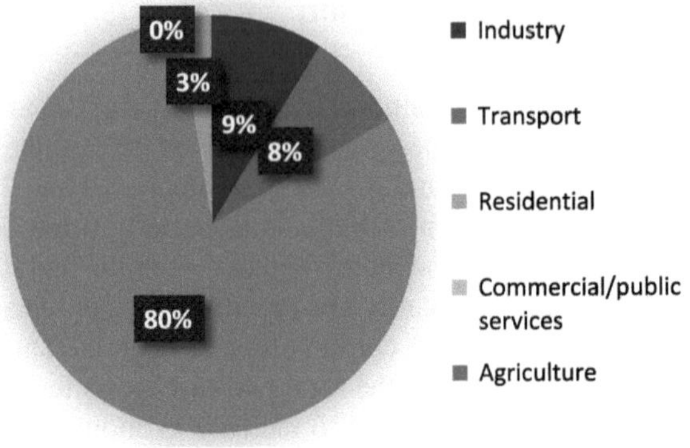

Fig. 2.23 Total primary energy consumption in Nigeria by sector. *Source* IEA (2015b)

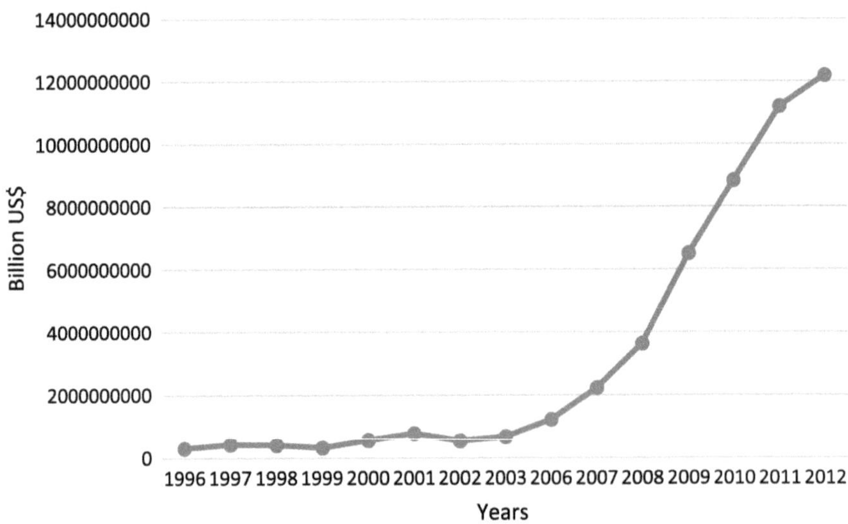

Fig. 2.24 Automobile import in Nigeria. *Source* NCS (2015)

As shown in Fig. 2.23, the highest final energy consumption in the various sectors in Nigeria is the residential sector with 80 %, while the industrial and agricultural sectors account for 9 and 8 % respectively. With the rapidly increasing numbers of vehicle import in Nigeria (see Fig. 2.24), the rate of energy consumption is expected to increase in the transportation sector just as in other sectors of the economy. This growth with the number of imported vehicles was on the increase from 2006 (about US$1.2 billion) to 2012 (about US$12.2 billion).

The rate of utilization of oil products such as petrol and diesel in the transportation sector is the highest in the total energy consumption as compared to other sectors such as the industrial, commercial/public services, and residential sectors

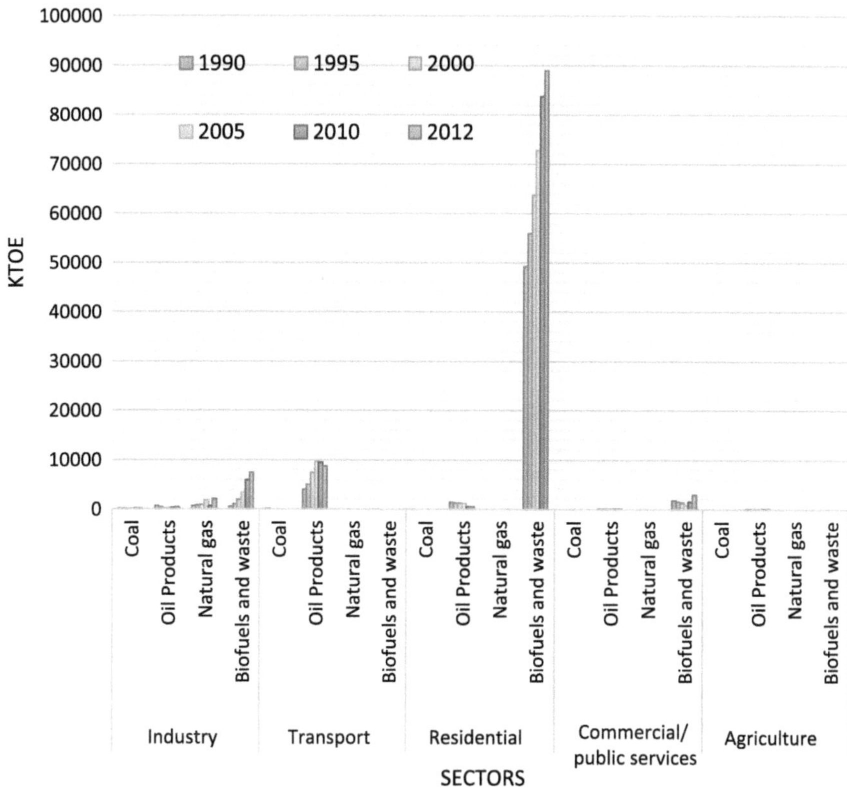

Fig. 2.25 Total primary energy consumption by fuel and sectors. *Source* IEA (2015b)

which mainly use diesel for electricity generation. This is observed in Fig. 2.25, which shows the combination of primary energy consumption by both fuel and sectors.

To address the issue of desertification in Nigeria, the Nigerian Government prepared the National Report titled, "Combating Desertification and Mitigating the Effects of Drought in Nigeria", which contains various strategic policies (FRN 1999). A more recent strategic plan by the Nigerian Government was the "National Strategic Action Plan" which was developed in 2012 by the Ministry of Environment (FME 2012). These documents as developed by policy makers face some implementation challenges such as the provision of alternative source of fuel other than firewood, since the population of the rural population is expected to rise. There is also the issue of urbanization as rural dwellers will migrate to urban areas when there is no longer firewood and diesel is readily available in the urban areas.

On electricity consumption in Nigeria (Fig. 2.26), the growing trend of energy consumption rate is also observed for electricity consumption in the residential sector, from 3,949 GWh in 1990 to 14549 GWh in 2012. The commercial/ public services electricity consumption was 6,627 GWh in 2012, while the industrial electricity consumption was 3,983 GWh. Considering Fig. 2.26, electricity

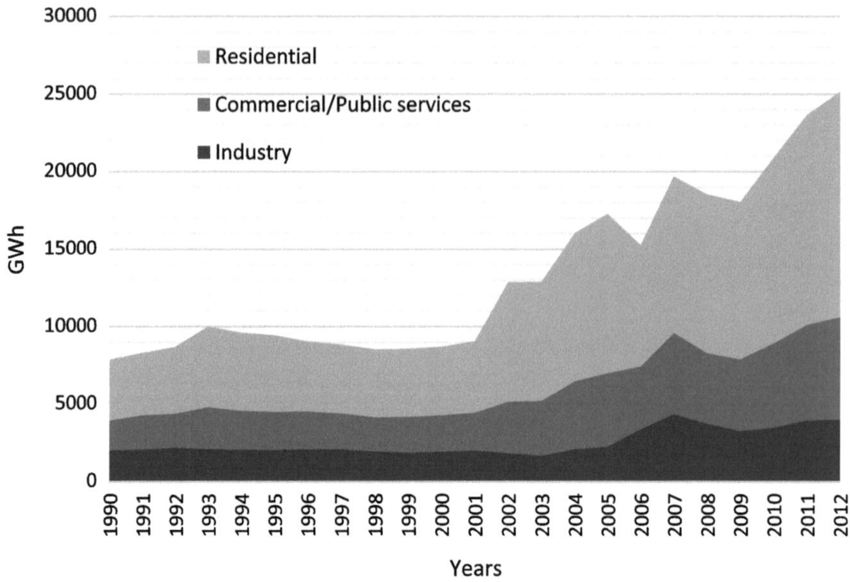

Fig. 2.26 Electricity consumption by sector. *Source* IEA (2015c)

consumption in the industrial sector did not increase much in 2012, as compared to 1990. This low electricity has forced most industries to depend on private generators to run their operations and businesses (Emodi and Boo 2015a). The South African mobile phone company MTN who operates the largest mobile phone company in Nigeria is estimated to have installed 6,000 generators to supply its base stations for up to 19 h a day. This cost the company $5.5 million a month on diesel in order to run the generators (Lawal 2007).

2.5 Relevant Government Ministries, Parastatals and Agencies in the Nigerian Energy Sector and Their Roles

The relevant government ministries, parastatals, and agencies are presented as follows.

2.5.1 The Energy Commission of Nigeria (ECN)

The Energy Commission of Nigeria (ECN) was established by Act No. 62, as amended by Act No. 32 of 1988 and Act No. 19 of 1989. It commenced operation in 1989 and is in charge of the energy sector planning and policy implementation.

The ECN promotes the use of renewable and alternative energies in the country's energy mix, as well as fulfils the role of strategic overall planning, coordination and effective direction of Nigeria's national energy policies and strategies. The Commission (i.e. ECN) has the following responsibilities (ECN 2015):

- To serve as a center for gathering and dissemination of information relating to national policy in the field of energy;
- To serve as a center for solving any inter-related technical problems that may arise in the implementation of any policy relating to the field of energy;
- To advise the government of the Federation or a State on questions relating to such aspect of energy as the Government of the Federation or a State may from time to time refer to it;
- To prepare after consultation with such agencies of government whose functions relate to the field of energy development or supply as the Commission considered appropriate, periodic master plans for the balanced and coordinated development of energy in Nigeria and such plans include:
  - Recommendations for the exploitation of new sources of energy as and when considered necessary, and
  - Such other recommendations to the Government of the Federation relating to its functions under this Decree as the Commission may consider to be in the national interest;
- To lay down guidelines on the utilization of energy types for specific purposes;
- To inquire into and advise the Government of the Federation or of the State on the adequate funding of the energy sector including research and development, production and distribution;
- To collate, analyze, and publish information relating to the field of energy;
- To carry out such other activities as are conclusive to the discharge of its functions under this Decree;
- To monitor the performance of the energy sector in the execution of government policies on energy;
- To promote training and manpower development in the energy sector;
- To liaise with all international organizations in energy matters.

2.5.2 Federal Ministry of Environment (FME)

The Federal Ministry of Environment (FME) was established in 1999 by the Federal Government of Nigeria with the statutory responsibility of protecting the natural environment against pollution, degradation, and conservation of natural resources. The Ministry (i.e. FME) is officially charged with environmental responsibilities and also with coordinating all climate change matters under its Special Unit Climate Change (SUCC). The unit also represents the Ministry at international climate negotiations. The Special Unit on Renewable Energy (SURE) acts as the voice of the ministry regarding renewable energy and energy efficiency issues. The objective of the unit is to develop and implement strategies that will achieve clean, reliable energy supply mechanisms in order to develop the sector based on international best practices and in order to showcase viability for private sector participation. The official mandates of the FME are presented as follows (FME 2015);

- To prepare a comprehensive National Policy for the protection of the environment and conservation of natural resources, including procedure for environmental impact assessment of all developing projects.
- To prepare in accordance with the National Policy on Environment, periodic master plans for redevelopment of environmental science and technology and advise the Federal Government on the financial requirements for the implementation of such plans.
- To advise the Federal Government on National Environmental Policies and priorities, the conservation of natural resources and sustainable development and scientific and technological activities affecting the environment and natural resources.
- To promote cooperation in environmental science and conservation technology with similar bodies in other countries and with international bodies connected with the protection of the environment and the conservation of natural resources.
- To cooperate with the Federal and State Ministries, Local Government, statutory bodies, and research agencies on matters and facilities relating to the protection of the environment and the conservation of natural resources,
- To prescribe standards for and make regulations on water quality, effluent limitations, air quality, atmospheric protection, ozone protection, noise control as well as the removal and control of hazardous substances, and
- To monitor and enforce environmental protection measures.

2.5.3 Federal Ministry of Lands, Housing, and Urban Development (FMLHUD)

The Federal Ministry of Lands Housing and Urban Development (FMLHUD) is an important arm of the Nigerian government that is charged with the implementation of all policies and regulations relating to lands, housing and urban development in the country. The FMLHUD has the mandate to enforce regulations in buildings in all the sectors of the economy. This involves the ministry's strategic role in energy efficiency in the residential, industrial, as well as commercial/services sectors. This is carried out under the Architectural Services under the FMLHUD and the inclusion of energy efficiency under the newly developed building code. Other mandates of the FMLHUD are the establishment of sustainable housing delivery system which ensures easy access to home ownership and rental schemes by the Nigerian populace in an environment where basic physical and social amenities are available (FMLHUD 2015).

2.5.4 Federal Ministry of Power (FMP)

The Federal Ministry of Power has the official responsibility of ensuring a robust power sector that fully supports the socio-economic needs of the nation. The main goal of the ministry is directed at initiating, formulating, coordinating, and implementing broad policies and programmes on the development of electricity generation from all sources of energy. FMP is also charged with the responsibility of developing and deploying electricity-related renewable energy policies in Nigeria. The FMP also ensures that Nigeria is provided with adequate and reliable power supply by implementing generation, transmission, and distribution projects in the sector and facilitates the emergence of a private sector led competitive and efficient power industry. In a view to diversify the nation's energy mix, the FMP encourages the utilization of renewable energy for power generation in both the urban and rural areas (FMP 2015).

2.5.5 Federal Ministry of Water Resources (FMWR)

The Federal Ministry of Water Resources (FMWR) was created in its current form in April 2010 with the responsibility of providing sustainable access to safe and sufficient water to meet the socio-economic needs of all Nigerians. This is achieved through efficient water resources management for basic human needs, irrigated agriculture, hydropower generation, and the promotion of healthy population while maintaining the integrity of fresh water bodies. The FMWR is involved in numerous renewable energy and rural electrification activities through its Department of Dams and Reservoir Operations. To increase energy supply in order to meet the nation's energy demands as mandated by the government, FMWR is collaborating with the FMP on the aspect of handling the area of power generation component, while the ministry handles the aspect of civil works in all the dam projects with hydropower potential. Small hydropower schemes have been integrated into some dam projects across the country in order to increase the energy supply of the nation (FMWR 2015).

2.5.6 National Power Training Institute of Nigeria (NAPTIN)

The National Power Training Institute of Nigeria (NAPTIN) was established on March 23rd 2009 and commenced full operation in September, 2009. The primary purpose of its establishment is to provide training for power sector personnel and coordinate training activities in the sector. In pursuit of this mandate, NAPTIN has taken over the management of existing seven regional training centers of PHCN. NAPTIN reports directly to the FMP and it is supervised by a board of directors appointed by them (NAPTIN 2015).

2.5.7 Nigerian Bulk Electricity Trading Plc. (NBET)

In 2001, the Federal Government, in a bid to address the deficiency in power sector supply, adopted the National Electric Power Policy for the reform of the sector. Following suit, the Electric Power Sector Reform Act (EPSRA), was passed into law in March 2005. A key thrust of the sector reform is the transfer of the control and operations of the industry from public sector to private sector. In furtherance of this goal, the Act saw the creation of the Power Holding Company Nigeria (PHCN) which assumed the assets, liabilities, and employees of the erstwhile Nigeria Electricity Power Authority (NEPA); the subsequent unbundling of PHCN into 18 successor companies, the establishment of the Nigerian Electricity Regulatory Commission (NERC); the establishment of the Rural Electrification Agency (REA) and the provision for the establishment of two special purpose vehicles (SPVs) to undertake electric power trading and management of extant liabilities respectively.

In line with the "Road-map to Power Sector Reform" of August 2010, and, in fulfillment of the requirements of EPSRA, the Nigerian Bulk Electricity Trading Plc., (NBET) aka the Bulk Trader, was incorporated on July 29, 2010 as the SPV for carrying out, under license from National Electricity Regulatory Commission (NERC), the bulk purchase and resale function contemplated by the EPSRA. As such NBET has been set up to "engage in the purchase and resale of electric power and ancillary services from independent power producers and from the successor generation companies" (NBET 2015).

2.5.8 Nigerian Electricity Regulatory Commission (NERC)

The Nigerian Electricity Regulatory Commission was established as an independent regulatory agency in 2005 under the EPSR Act 2005. Its mandate is to monitor and regulate the electricity industry of Nigeria, and ensure compliance with market rules and operating guidelines. NERC is responsible for assessing applications for licenses to operate an independent power plant larger than 1 MW, and thus approves eligibility of prospective companies to negotiate a power purchase agreement with the central off-taker in the current transitional market, the NBET. The NERC also plays a crucial role in consumer protection by ensuring the development of customer service standards, fair pricing rules, and the provision of dispute resolution if a situation arises (NERC 2015).

2.5.9 Nigerian National Petroleum Corporation (NNPC)

The Nigerian National Petroleum Corporation (NNPC) is the corporation through which the federal government of Nigeria (the Department of Petroleum Resources, DPR) regulates and participates in the country's petroleum industry. NNPC was established on April 1, 1977 as a merger of the Nigerian National Oil Corporation and the Federal Ministry of Mines and Steel. NNPC by law manages the joint venture between the Nigerian federal government and a number of foreign multinational corporations, which include Royal Dutch Shell, Agip, ExxonMobil, Chevron, and Texaco (now merged with Chevron). Through collaboration with these companies, the Nigerian government conducts petroleum exploration and production.

NNPC has the sole responsibility for upstream and downstream developments, and is also charged with regulating and supervising the oil industry on behalf of the Nigerian Government. In 1988, the corporation was commercialized into 11 strategic business units, covering the entire spectrum of oil industry operations: exploration and production, gas development, refining, distribution, petrochemicals, engineering, and commercial investments (The Economist 2007; NNPC 2015c). The NNPC is also exploring the biofuel option for the transportation sector and this includes mainly ethanol and biodiesel which are mixed with conventional fuel.

2.5.10 Presidential Taskforce on Power (PTFP)

The Presidential Task Force on Power (PTFP) was established in June 2010, to drive the implementation of the reform of Nigeria's power sector. It brings together all the agencies that have a role to play in removing legal and regulatory obstacles to private sector investment in the power industry. It also has the mandate to monitor the planning and execution of various short-term projects in generation, transmission, distribution, and fuel-to-power that are critical to meeting the stated service delivery targets of the power reform roadmap (PTFP 2015).

The PTFP collaborates closely with various ministries and agencies that have specific contributions to the reform process, including the Federal Ministry of Power (FMP), the Federal Ministry of Finance (FMF), Ministry of Petroleum Resources (MPR), the Bureau of Public Enterprises (BPE), the Nigerian Electricity Regulatory Commission (NERC), the Nigerian National Petroleum Corporation (NNPC), the Bureau of Public Procurement, National Gas Company Limited (NGC) and the Power Holding Company of Nigeria (PHCN).

2.5.11 Rural Electrification Agency of Nigeria (REA)

The Rural Electrification Agency (REA) was established in 2006 under Section 88 (I) of the Electric Power Sector Reform Act (EPSRA), 2005. Its vision was to mobilize capital for sustainable private sector driven investment in rural electricity development in Nigeria dedicated to the goal of improving the living conditions in rural areas through enhancing agriculture, commercial, industrial, and domestic activities. The core functions of the REA are to promote and coordinate rural electrification projects, implement and manage the Rural Electrification Fund, and regulation of rural electrical sector. The REA projects center on grid expansion to rural areas via funding from the federal government's annual budgetary allocation; most recently it has broadened its ambit to include the deployment of off-grid and mini-grid renewable energy generating systems to accelerate the pace of improvement (REA 2015).

2.5.12 Standards Organization of Nigeria (SON)

Established by the Federal Government by the Act No. 56 of 1971, the Standards Organization of Nigeria (SON) is entrusted with the responsibility to ensure imported and manufactured products in Nigeria are kept at stipulated standards (SON 2015). Other functions of the SON include the following:

- To ensure the compliance of designated and approved standards by the Council;
- To undertake investigations as necessary into the quality of facilities, materials, and products in Nigeria, and establish a quality assurance system including certification of factories, products, and laboratories;
- To provide reference standards for calibration and verification of metering and metering equipment;
- To compile an inventory of products requiring standardization;
- To prepare Nigeria Industrial Standards;
- To foster interest in the recommendation and maintenance of acceptable standards by industry and the general public;
- To develop methods for testing of materials, supplies, and equipment;
- To register and regulate standard marks and specification.

Some renewable energy and energy efficiency standards have recently been developed and adopted by the SON. This includes among others, a code of practice for the deployment of outdoor solar lighting, design qualification and type approval of PV modules, safety standards for use of PV power converters, etc. (Ley et al. 2014).

2.5.13 Council for Renewable Energy in Nigeria (CREN)

The Council for Renewable Energy Nigeria is a multi-stakeholder association, which promotes the appropriate use of renewable energy technology in Nigeria and the reduction of greenhouse gases through reduced consumption of fossil fuels. CREN aims to bring together the professional sector, government and civil servants, academics, associations, industry, financial institutions and services, the not for profit sector, and end-users to act as a forum where they can work together for efficient, appropriate renewable energy implementation and to develop a comprehensive sustainable energy strategy for Nigeria.

The Council works with all stakeholders to address the challenges of awareness, availability, cost, and appropriate implementation of renewable energy technologies in Nigeria. Also, CREN creates public awareness and fosters the emerging availability of reliable, economically viable renewable energy systems by supporting policy information and implementation, research, development and use of such systems (CREN 2015).

2.5.14 *Green Building Council of Nigeria (GBCN)*

Although Nigeria is a prospective member of the World Green Building Council, the Green Building Council of Nigeria (GBCN) will aid in the development of green building rating system in Nigeria. The council will also provide a single useful metric for the establishment of green building actions and address the issue of climate change in Nigeria (WGBC 2015). Some rating systems that would be adopted include the following;

- BEREEAM (Building Research Establishment's Environment Assessment Method)
- CASBEE (Comprehensive Assessment System for Building Environmental Efficiency)
- SBTool (Sustainable Building Tool)
- Green Globes
- LEED (Leadership in Energy and Environmental Design).

2.6 Nigerian Energy Policies and Strategies

The Nigerian energy policies and strategies are summarized in this section from the earliest to the latest.

2.6.1 *National Electric Power Policy (NEPP), 2001*

The National Electric Power Policy (NEPP) was the first of its kind in the wake of reforms in the Nigerian power sector. Its development was due to the recommendations of the Electrical Power Implementation Committee (EPIC), which was the body in charge of reforms and transformation of the power sector in 1999.

The NEPP was created in March 2001,[24] and presented three bold steps in achieving the goal of reforming the power sector. The first step was to privatize NEPA which was state owned and introduce Integrated Power Producers (IPPs) of electricity. The next step was to increase competition between participants in the market, gradually remove subsidies and sell excess power to the DISCOs. In the last step, it was expected that the market and competition would have been more intense and allow for full cost-reflective pricing of supply, and liberalization of the electricity market would have been complete (Maduekwe 2011). The NEPP had the following as critical objectives for Nigeria's electric power sector:

- Ensure that the power sector attracts private investment both from Nigeria and abroad;
- Drafting of a new electricity law to provide the legal framework for the reform agenda;
- Establishment of an independent regulatory agency;
- Development of a wholesale electricity market;
- Establishment of a consumer assistance fund to ensure the efficient and targeted application of subsidies to less privileged Nigerians;
- Establishment of a Rural Electrification Agency ("REA") to manage the rural electrification fund.

2.6.2 National Energy Policy (NEP) (2003), (2006), (2013)

Before the Federal Government of Nigeria approved the energy policy in the year 2003, there was no comprehensive energy policy. The established energy policy was called the National Energy Policy (NEP) which was developed by the Energy Commission of Nigeria (ECN). The National Energy Policy (NEP) sets out government policy on the production, supply and consumption of energy reflecting the perspective of its overall needs and options.

The main goal of the policy is to create energy security through a robust energy supply mix by diversifying the energy supply and energy carriers based on the principle of "an energy economy in which modern renewable energy increases its share of energy consumed and provides affordable access to energy throughout Nigeria, thus contributing to sustainable development and environmental conservation". Importantly, the national policy already outlines the key elements for development and application of renewable energy as:

- To promote decentralized energy supply especially in rural areas, based on RE resources;
- To develop, promote, and harness Renewable Energy (RE) resources of the country and incorporate all viable ones into the national energy mix;

[24]See www.nigeriaelectricityprivatisation.com/?page_id=2.

- To promote efficient methods in the use of biomass energy resources;
- To de-emphasize and discourage the use of wood as fuel;
- To keep abreast of international developments in RE technologies and applications.

The NEP was first revised in 2006 to incorporate some changes and improvement, while the last draft revised edition was done in 2013. The 2013 draft revised edition includes environmental and climate change policy, policy on other energy issues such as RandD, bilateral, regional, and international cooperation, local content, manpower development and training, and Gender issues. Other areas included are policy on energy financing, planning, and policy implementation (see NEP 2003, 2006, 2013).

2.6.3 National Economic Empowerment and Development Strategy (NEEDS) (2004)

The National Economic Empowerment and Development Strategy (NEEDS) was developed by the National Planning Commission (NPA) in 2004 and was intended to develop and alleviate poverty in the country. This involves the action of human resources on the natural resources to produce goods necessary to satisfy the economic needs of the community.[25] On infrastructure, NEEDS promotes the privatization of government infrastructure and was one of the key instruments in achieving a revamped service delivery.

The Nigerian government will however, fund projects that have very low attractiveness and high investment cost to investors such as those in rural areas. Furthermore, the increased share of renewables in the national energy mix was further encouraged in the NEEDS. This involves the suggestion for the creation of renewable energy agency and technologies which will be funded under the National Power Sector Reform Act. This was a milestone towards the adoption of renewables in the power sector and their utilization for rural electrification (NEEDS 2004; Marcellus 2009).

2.6.4 National Power Sector Reform Act (EPSRA), 2005

The National Power Sector Reform Act established in 2005 ensured the liberalization of the Nigerian power sector. The Act was due to the NEPP developed in 2001 and made provision for new legal and regulatory framework for the power

[25]www.omojuwa.com/2014/06/national-economic-empowerment-and-development-strategy-needs%E2%80%8F-ayorinsola-obisanya/.

sector.[26] The Act gave way to unbundling and privatization of the power sector, which intends to introduce competition in the electricity market, enhance rural electrification, while protecting consumer rights and developing performance standards in the power sector.[27] The main provisions of the Act include the following:

- Creation of the initial Holding Company (PHCN) to assume the assets, liabilities and employees of NEPA;
- Unbundling of PHCN into successor companies and ensuring greater operational autonomy;
- Market development;
- Privatization of successor companies which empowers the Bureau to undertake this responsibility;
- Establishment of the Nigeria Electricity Regulatory Commission (NERC);
- Establishment of the Rural Electrification Agency and Fund to provide access to electricity to the rural areas as well as fund rural electrification projects, respectively;
- Power consumer assistance fund to bridge the funding gaps for low income earners.

The Act also provides for an investment-friendly environment for potential investors in the power sector by transforming the integrated structure of the sector (as in the period of NEPA) into vertically unbundled segments: generation, transmission and distribution (Ajumogobia and Okeke 2015). The Act ensured a transition in the reform process, providing necessary criteria for the advancement of power generation reforms. As part of the process in the Act, a wider mandate for power market regulation was established (Ayanruoh 2012).

2.6.5 Renewable Electricity Policy Guidelines (REPG) 2006

Developed by the federal Ministry of Power and Steel in December, 2006, the Renewable Electricity Policy Guidelines (REPG) mandated the Nigerian government on the expansion of electricity generation[28] from renewables to at least 5 % of the total electricity generated and a minimum of 5 TWh of electricity generation in the country (REPG 2006). This policy document presents the Nigerian government's plans, policies, strategies, and objectives for the promotion of

[26]www.reegle.info/policy-and-regulatory-overviews/ng.

[27]www.nigeriapowerreform.org/index.php?option=com_content&view=article&id=79:policy-framework&catid=41&Itemid=304.

[28]www.businessdayonline.com/2015/02/reviewing-the-legal-framework-for-renewable-energy-projects-in-nigeria/.

renewables in the power sector (Iwayemi et al. 2014). The policy goals and strategies of the REPG are as follows:

- Expansion of the market for renewable electricity to at least five percent of total electricity generating capacity and a minimum of 5 TWh of electric power production;
- Establishment of stable and long-term favorable pricing mechanisms and ensuring unhindered access to the grid with guaranteed purchase and transmission of all electricity produced by renewable electricity producers and obliging the grid operators upgrade the system accordingly;
- Construction of independent renewable electricity systems in areas not covered by the electricity grid;
- Development of innovative, cost-effective and practical measures to accelerate access to electricity services in rural areas through renewable sources;
- Setting up of a Renewable Electricity Trust Fund to be governed by the Rural Electrification Fund;
- Creation of a multi-stakeholder partnership for the delivery of renewable electricity to meet national development goals;
- Broadening international cooperation in expanding the role of renewable electricity for meeting national development goals and contributing to global efforts in addressing climate change.

2.6.6 *Renewable Electricity Action Programme (REAP) (2006)*

Developed in relation to the REPG by the Federal Ministry of Power and Steel in 2006, the Renewable Electricity Action Programme (REAP) set out a roadmap for the implementation of the REPG. The document presents an overview of the Nigerian electricity sector and relates it to renewable energy development. The documents also reviews government targets and provides strategies for renewable energy development such as leveling the playing field for renewable electricity producers, multi-sector partnerships, demonstration projects, supply chain initiatives, etc. The study also made provision for financing renewable programs and explored the roles of government ministries and agencies, then concludes with risk assessment, monitoring, and evaluation (REAP 2006). The Ministry of Power and Steel has however been restructured to the Ministry of Power and this programme may have been abandoned.

2.6.7 *Nigerian Biofuel Policy and Incentives (NBPI) (2007)*

The aim of this policy was to develop and promote the domestic fuel ethanol industry through the utilization of agricultural products. This was in line with the government's directive on an Automotive Biomass Programme for Nigeria in August

2005. The NNPC was mandated to create an environment for the take-off of the ethanol industry. The policy further aimed at the gradual reduction of the nation's dependence on imported gasoline, reduction in environmental pollution, while at the same time creating a commercially viable industry that can precipitate sustainable domestic jobs. The benefits of this policy were to create additional tax revenue, provision of jobs to reduce poverty, boost economic development and empower those in the rural areas, improve agricultural activities, energy and environmental benefits through the reduction of fossil fuel related GHGs in the transport sector.

The targets of the NBPI are as follows:

- To ensure the contribution of all biofuel companies with 0.25 % of their revenue towards funding research into feedstock production, local technology development and improved farming practice;
- To develop an import duty waiver for biofuels granted for 10 years;
- An exemption from taxation, withholding tax and capital gains tax imposed in respect of interest on foreign loans, dividends, services rendered from outside Nigeria to biofuel companies by foreigners;
- To launch a special kind of loan for investors in the biofuel industry to aid the development of large-scale schemes and large-scale integrated operation including plantation, a plant and within-the-gate collocated power generating plants;
- To achieve the blending of up to 10 % of fuel ethanol with gasoline to achieve a blend to be known as E-10 during the seeding phase of the programme;
- To achieve 100 % domestic production of biofuels consumed in the country by 2020;
- To ensure an off-take agreement by NNPC for biofuels as buyer of last resort.

The implementation of the NBPI was set to be undertaken in two phases. The first phase is called seeding the market which involves the importation of cargoes of fuel ethanol until the domestic market is matured. The seeding phase was expected to begin with initial penetration of selected cities during the first 3 years, while the nationwide roll-out is expected within 5–10 years. Phase 2 which is called biofuel production programme, will begin concurrently with the seeding programme. It was the core agriculture integration programme and ensures the establishment of plantations and the construction of a bio-fuel distillers and plants in Nigeria (NBPI 2007).

2.6.8 Nigerian Gas Master Plan (NGMP) 2008

In line with the Nigerian government's plan to become a major international player in the international gas market as well as to lay a solid framework gas infrastructure expansion within the domestic market, the Nigerian Gas Master Plan (NGMP) was developed and approved on the February13th, 2008.[29] The NGMP

[29]See www.nercng.org/index.php/myto-2.

envisages a wholesale transition to decentralized privately held electricity generation gas plants from the erstwhile public power utility. The plan also aims to stem the huge waste associated with gas flaring and to put to more productive use the nation's large gas reserves. What the Gas Master Plan does not clearly mention is that the utilization of compressed natural gas (CNG) can play a major role in transforming the nation's transportation sector if adopted on a large scale (Emodi and Boo 2015c).

2.6.9 Roadmap for Power Sector Reform (A Customer-Driven Sector-Wide Plan to Achieve Stable Power Supply 2010) and (Revision 1, 2013)

The first Roadmap for Power Sector Reform was developed by the Presidential Task Force on Power (PTFP) in 2010 and based on it, the revision was carried out and presented on the August 30th, 2013. This document includes review and strategic plans by the government to finalize the reforms in the power sector, while setting the nation on a steady path of clean electricity generation at competitive rates. The Roadmap is not in itself a policy document but a set of strategies to fast-track the achievement of the NEP (2003) as stated in the EPSRA of 2005. The Roadmap explores some key aspects of the power sector and development such as generation, transmission, distribution, NIPP and fuel supply to power plants (see RPSR 2010, 2013). Some recommendations and proposals of the reforms are given below:

- Improvement of collection efficiency and returns to the market during the pre-Transitional Electricity Market (TEM) declaration stage;
- The commencement and conclusion of all labor negotiations, settlement of liabilities, rationalization and eventual winding down of the Power Holding Company of Nigeria (PHCN);
- The continuation of the clear and firm political will to resist efforts that could undermine privatization and the reform;
- The development of an optimal transmission capacity expansion plan and funding strategy to provide a reliable highway for wheeling generated power;
- Acceleration of the management and operational efficiency levels of the Transmission Company of Nigeria (TCN);
- More commitment to deliver the gas development and transportation infrastructure projects earmarked for gas-to-power alignment;
- Prevention of frequent acts of vandalism to pipelines such as the Trans-Forcados and Trans-Niger crude oil lines, and the Escravos Lagos Pipeline Systems (ELPS) gas pipelines.
- Clarity on the interim operation and maintenance of the National Integrated Power Projects (NIPP) generation assets;

- More firm commitment from NIPP to deliver its critical transmission projects scheduled for 2013 as well as the Omoku, Gbarain, and Alaoji power plants to come on stream without further delay;
- Meeting of the conditions precedent to the declaration of TEM, before handing over the successor companies to the new owners.
- Positioning a well-capitalized Nigerian Electricity Liability Management Company (NELMCO) to address post hand-over fall outs from creditors.
- To secure a minimum, transitional service-delivery level through project and process optimization.

2.6.10 Renewable Energy Master Plan (REMP) (2005), (2012)

The Renewable Energy Master Plan (REMP) was developed by the Energy Commission of Nigeria (ECN), in collaboration with the United Nations Development Programme (UNDP) in 2005 and was later reviewed in 2012 (REMP 2005, 2012). The REMP expresses Nigeria's vision and sets out a road map for increasing the role of renewable energy in achieving sustainable development. The REMP is anchored on the mounting convergence of values, principles and targets as embedded in the National Economic Empowerment and Development Strategy (NEEDS), National Energy Policy, National Policy on Integrated Rural Development, the Millennium Development Goals (MDGs), and international conventions to reduce poverty and reverse global environmental change (REMP 2012).

The REMP stresses the need for the integration of renewables in buildings, electricity grids, and for off-grid electrical systems. Further, the importance of solar power in the country's energy mix was also highlighted in the policy document. According to the REMP, Nigeria intends to increase the supply of renewable electricity from 13 % of total electricity generation in 2015 to 23 % in 2025 and 36 % by 2030. Renewable electricity would then account for 10 % of Nigeria's total energy consumption by 2025. However, the REMP has not been approved by the National Assembly to be passed into law.

2.6.11 National Renewable Energy and Energy Efficiency Policy (NREEEP) (2014)

The National Renewable Energy and Energy Efficiency Policy (NREEEP) outlines the global thrust of the policies and measures for the promotion of renewable energy and energy efficiency. The FMP developed the NREEEP in 2014 and is awaiting the approval of the Federal Executive Council. The objectives of the NREEEP are presented as follows (NREEEP 2014);

- To set out a framework for action to address Nigerians' challenge of inclusive access to modern and clean energy resources, improved energy security, and climate objectives;
- To recognize the national significance of renewable electricity generation activities by providing for the development, operation and maintenance, and upgrading of new and existing renewable electricity generation activities;
- To declare that the proportion of Nigeria's electricity generated from renewable energy sources shall increase to a level that meets or exceeds the ECOWAS regional policy targets for renewable electricity generation and energy efficiency for 2020 and beyond;
- To declare energy efficiency to be a major, low-cost, and under-utilized Nigerian energy resource offering savings on energy bills, opportunities for more jobs, improving industrial competitiveness, and lowering air pollution;
- To recognize that poverty mitigation and environmental protection are hindered by the continued predominance and inefficient use of oil and natural gas in meeting our energy needs;
- To take a step in the right direction and broaden the definition of energy security to include renewable energy and energy efficiency as equally important indigenous sources of energy, in addition to oil and gas;
- To incorporate provisions for renewable energy and energy efficiency generation activities into government policy statements and plans, and recognize the importance of enabling framework conditions for private investment in renewable energy and energy efficiency;
- To set national targets for achievements in electricity from renewable energy and energy efficiency capacity addition by 2020 and beyond;
- To require the preparation of a national action plan for renewable energy and for energy efficiency, and set a time frame within which implementation is required;
- To recommend that the signatory parties to this policy should collaborate in preparation of the action plans and work together in achievement of the final mandatory targets;
- To make it mandatory for the Ministry of Power to facilitate the development of an integrated resource plan (IRP) and ensure the continuous monitoring and review of the implementation and effectiveness of the action plans prescribed under the national policy statement;
- To take steps away from the overheated rhetoric that Nigeria's future energy independence be secured by ever more gas and oil consumption; and
- To facilitate the establishment of a framework for sustainable financing of renewable energy and energy efficiency projects and programmes in Nigeria.

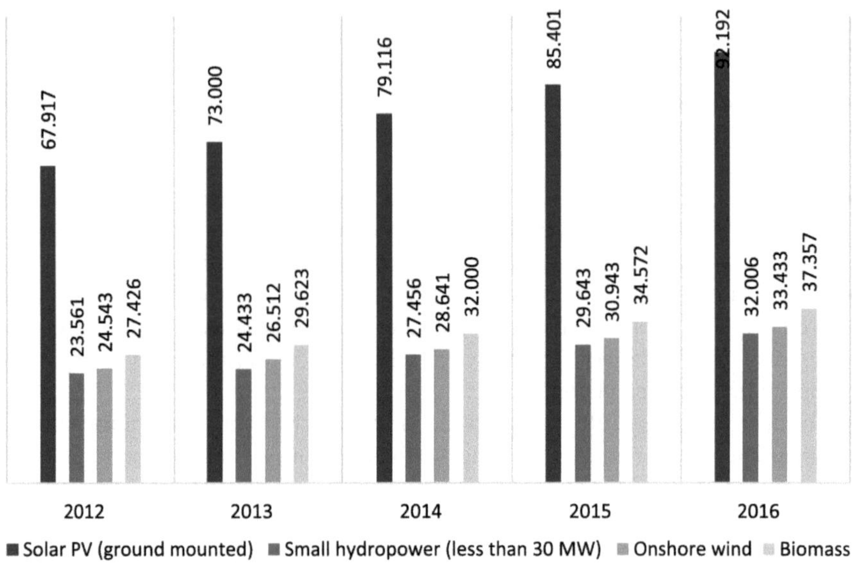

Fig. 2.27 Nigeria's MYTO 2 FiTs for 2012–2016 (N/Per MWh)

2.6.12 Multi-year Tariff Order (MYTO), 2008 and 2012

In 2008, a 15-year roadmap towards cost-reflective tariffs called the Multi-Year Tariff Order (MYTO 1) was developed by the Nigerian Electricity Regulatory Commission (NERC). The first two phases, 2008–2000 and 2012–2017, were designed to keep consumer prices relatively low, through affecting the price increases in a gradual manner. The final regime is intended to provide the necessary incentives for power producers and investors to operate and maintain electricity infrastructure (Emodi and Yusuf 2015a). The NERC has released the Multi-Year Tariff Order 2 (MYTO 2), which has similar features to MYTO 1 but includes some improvements, and will be effective from June 1st 2012 to May 31st 2017. The retail tariff in MYTO 2 will be reviewed bi-annually and changes may be made for all electricity generated at wholesale contract prices, adjusted for the Nigerian inflation rate, US$ exchange rate, daily generation capacity, and accompanying actual CapEx and OpEx requirements that will vary from those used in the tariff calculation. Figure 2.27 shows the MYTO 2 feed-in tariffs from 2012 to 2016 in Nigeria (Nigerian Naira per MWh (N/MWh) and the dollar equivalent).[30]

The review of all inputs to the tariff calculation is expected to begin by 2016 as the basis of a new Multi-Year Tariff Order (MYTO) designed to kick-start the next five years starting from June 1st 2017. The MYTO 2 contains a 15-year

[30]www.nercng.org/index.php/myto-2.

tariff pathway for electricity generated from RE, with bi-annual minor reviews and major reviews every five years. The MYTO 2 tariffs are negotiable if a generator can prove to the NERC that their costs for electricity generation from renewables are not in line with the assumptions of the MYTO 2 (Emodi and Boo 2015c).

2.6.13 *Draft Rural Electrification Strategy and Implementation Plan (RESIP) (2014)*

The Power Sector Reform team initially prepared the Rural Electrification Strategy and Implementation Plan in 2006 (RESIP 2006). However, a committee involved in the power sector reviewed and redrafted the RESIP in 2014. It was expected to establish a clear institutional step-up for the sector and set a roadmap which will result in the development of an enabling framework for rural electrification in Nigeria. The primary objective of the RESIP is to expand access to electricity as rapidly as can be afforded in a cost-effective manner. This includes the use of on-grid and off-grid means of electricity supply. The draft is ready and awaiting approval from the government. A general overview of the various Nigerian energy policies, legislations, regulations, standards, programmes and incentives, their achievements, gaps, alignment, overlaps, and opportunities is presented in Fig. 2.28.

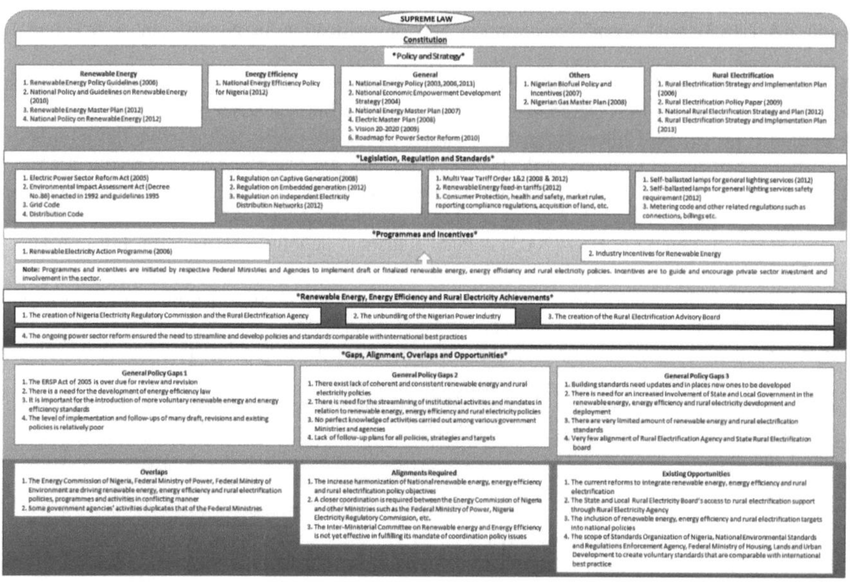

Fig. 2.28 An overview of the Nigerian energy policies. *Source* Ley et al. (2014), Energypedia (2015)

References

Adaramola MS, Oyewola OM (2011) Wind speed distribution and characteristics in Nigeria. ARPN J Eng Appl Sci 6(2):82–86

Adekoya LO, Adewale AA (1992) Wind energy potential of Nigeria. Renew Energy 2(1):35–39

Adeoti O (1998) Engineering economy studies of biogas as a renewable energy source at household level in Nigeria. Technology Planning and Development Unit, Faculty of Technology, Obafemi Awolowo University, Ile-Ife, Nigeria

Ajao KR, Ajimotokan HA, Popoola OT, Akande HF (2009) Electric energy supply in Nigeria, decentralized energy approach. Cogen Distrib Gen J 24(4):34–50

Ajumogobia HO, Okeke CN (2015) Nigerian energy sector: legal and regulatory overview. Ajumogobia and Okeke. www.ajumogobiaokeke.com/media/publications/nigerian-energy-sector-legal-regulatory-overview-2015. Accessed 6 Aug 2015

Akinbami JF, Akinwumi I, Salami A (1996) National resources forum. UN Sustain Dev J 20(4):319–332

Akinbami JF, Ilori MO, Oyebisi TO, Akinwumi IO, Adeoti O (2001) Biogas energy use in Nigeria: current status, future prospects and policy implications. Renew Sustain Energy Rev 5(1):97–112

Akpan U (2015) Technology options for increasing electricity access in areas with low electricity access rate in Nigeria. Socio-Econ Plann Sci 51:1–12

Akubo S, Dongo EI, Momoh IM, Okorie NN, Oluyori RT (2013) Revitalization of the Nigerian coal mining industry to expand the power generation needs of Nigeria. J Res Environ Sci Toxicol 2(8):175–178

Aladejare SA (2014) Energy, growth and economic development: a case study of the Nigerian electricity sector. Am J Bus Econ Manage 2(2):41

Al-Amin M (2014) Domestic energy crisis and deforestation challenges in Nigeria. J Environ Earth Sci 4(2):94–100

Aliyu UO, Elegba SB (1990) Prospects for small hydropower development for rural applications in Nigeria. Nigerian J Renew Energy 1(1):74–86

Audu EB (2013) Fuel wood consumption and desertification in Nigeria. Int J Sci Technol 3(1):1–5

Augustine C, Nnabuchi MN (2009) Relationship between global solar radiation and sunshine hours for Calabar, Port Harcourt and Enugu, Nigeria. Int J Phys Sci 4(4):182–188

Ayanruoh F (2012) Will liberalizing the Nigerian electricity power sector solve the problem of shortage of capacity? www.works.bepress.com/felix_ayanruoh/2. Accessed 6 Aug 2015

Balogun AZ (2010) Gas to power: the impact of natural gas pricing on electricity tariffs and supplier benefits in Nigeria. M.Sc. thesis. University of Oklahoma

Central Bank of Nigeria (CBN) (2005) Annual statistical bulletin, vol 16. CBN Press, Abuja, p 2005

Central Bank of Nigeria (CBN) (2015a) Nigeria's domestic crude oil production and export. www.cenbank.org/rates/crudeoil.asp. Accessed 2 Jan 2015

Central Bank of Nigeria (CBN) (2015b) Crude oil prices in Nigeria. www.cenbank.org/rates/crudeoil.asp. Accessed 2 Jan 2015

Central Intelligence Agency (CIA) (2015a) Dry natural gas production and consumption in Nigeria. www.cia.gov. Accessed 3 Jan 2015

Central Intelligence Agency (CIA) (2015b) Coal production in Nigeria. www.cia.gov. Accessed 8 Jan 2015

Council for Renewable Energy Nigeria (CREN) (2015) http://www.renewablenigeria.org.ng/index.php/en/. Accessed 6 Aug 2015

Detail Commercial Solicitors (2014) The declaration of transitional electricity market (TEM) and stakeholder expectations for the Nigerian electricity supply industry (NESI). www.detailsolicitors.com/?section=news&cmd=details&newsid=49&printview=1. Accessed 30 Dec 2014

Detail Commercial Solicitors (2015) Nigeria power guide, vol 3. www.lexology.com/library/detail.aspx?g=f77e24d6-8338-47d6-9e4d-ed5b89a51862. Accessed 30 Dec 2015
Eludoyin OM (2015) The challenge of reducing food carbon footprint in a developing country. J Clim Weather Forecast 3(124):2
Emodi NV, Boo KJ (2015a) Sustainable energy development in Nigeria: current status and policy options. Renew Sustain Energy Rev 51:356–381
Emodi NV, Boo KJ (2015b) Decomposition analysis of CO2 emissions from electricity generation in Nigeria. Int J Energy Econ Policy 5(2):565–573
Emodi NV, Yusuf SD (2015) Improving electricity access in Nigeria: obstacles and the way forward. Int J Energy Econ Policy 5(1):335–351
Energy Commission of Nigeria (ECN) (2007) National energy policy. Federal Republic of Nigeria, Abuja 2007
Energy Commission of Nigeria (ECN) (2009) Renewable energy master plan. www.energy.gov.ng. Accessed 8 Jan 2015
Energy Commission of Nigeria (ECN) (2015) www.energy.gov.ng. Accessed 6 Aug 2015
Energy Intelligence (2015) Nigeria's crude oil. www.energyintel.com/pages/eigarticle.aspx?DocId=200017. Accessed 1 Dec 2015
Energypedia (2015) Nigeria energy situation. Energypedia. www.energypedia.info/wiki/Nigeria_Energy_Situation#cite_note-Nigerian_Energy_Support_Programme.2C_To_be_published.2C_The_Nigerian_Energy_Sector_-_an_Overview_with_a_Special_Emphasis_on_Renewable_Energy.2C_Energy_Efficiency_and_Rural_Electrification-16. Accessed 6 Aug 2015
Enibe SO, Odukwe AO (1990) Patterns of energy consumption in Nigeria. Energy Convers Manag 30(2):69–73
Essien AU, Igweonu EI (2014) Coal based generation: a solution to Nigeria electricity problem. Int Arch Appl Sci Technol 5(1):74–80
Fagbenle RO, Karayiannis TG (1994) On the wind energy resources of Nigeria. Int J Energy Res 18(5):493–508
Federal Ministry of Environment (FME) (2012) Great Green Wall for the Sahara and Sahel initiative: national strategic action plan. www.fao.org/fileadmin/templates/great_green_wall/docs/NIGERIA-GGWSAP__FINAL_Oct_2012.pdf. Accessed 6 April 2015
Federal Ministry of Environment (FME) (2015) www.environment.gov.ng. Accessed 6 Aug 2015
Federal Ministry of land, Housing and Urban Development (FMLHUD) (2015) www.landsandhousing.gov.ng. Accessed 6 Aug 2015
Federal Ministry of Power (FMP) (2015) www.power.gov.ng. Accessed 6 Aug 2015
Federal Ministry of Water Resources (FMWR) (2015) www.waterresources.gov.ng. Accessed 6 Aug 2015
Felix AA, Akinbulire TO, Abdulkareem A, Awosope COA (2012) Wind energy potential in Nigeria. Int Electr Eng J (IEEJ) 3(1):595–601
Garba B, Bashir AM (2002) Managing energy resources in Nigeria: studies on energy consumption pattern in selected rural areas in Sokoto state. Nigerian J Renew Energy 10(1&2):97–107
Hui SCM (1997) From renewable energy to sustainability: the challenge for Honk Kong. Hong Kong Institution of Engineers, pp 351–358
Ibitoye FI (2014) Ending natural gas flaring in Nigeria's oil fields. J Sustain Dev 7(3):13
International Energy Agency (IEA) (2015a) Nigeria's total primary energy supply (1990–2012). www.iea.org/stats/WebGraphs/NIGERIA5.pdf. Accessed 6 Jan 2015
International Energy Agency (IEA) (2015b) Primary energy consumption in Nigeria by fuel. www.iea.org/statistics/statisticssearch/report/?country=NIGERIA&product=Balances&year=2012. Accessed 6 Jan 2015
International Energy Agency (IEA) (2015c) Electricity consumption by sector in Nigeria (1990–2012). http://www.iea.org/statistics/statisticssearch/report/?country=NIGERIA&product=electricityandheat&year=2012. Accessed 9 May 2015

Iwayemi A, Diji C, Awotide B, Adenikinju A, Obute P (2014) Towards sustainable universal electricity access in Nigeria. CPEEL. www.cpeel.ui.edu.ng/sites/default/files/monograph-2.pdf. Accessed 6 Aug 2015

Kennedy-Darling J, Hoyt N, Murao K, Ross A (2008) The energy crisis of Nigeria: an overview and implications for the future. The University of Chicago, Chicago

KPMG (2013) A guide to the Nigerian power sector. www.kpmg.com/Africa/en/IssuesAndInsights/Articles-Publications/Documents/Guide%20to%20the%20Nigerian%20Power%20Sector.pdf. Accessed 30 Dec 2014

Lawal L (2007a) Lights out for oil-rich Nigeria. Fortune-Energy Shortages. www.archive.fortune.com/2007/12/03/news/international/nigeria_power.fortune/index.htm?section=money_news_international

Lawal L (2007b) Lights out for oil-rich Nigeria. Fortune-Energy Shortages

Ley K, Gaines J, Ghatikar A (2014) The Nigerian energy sector—an overview with a special emphasis on renewable energy, energy efficiency and rural electrification. GOPA International Energy Consultants GmbH, Germany. www.giz.de/en/downloads/giz2014-en-nigerian-energy-sector.pdf. Accessed 6 Aug 2015

Maduekwe NC (2011) Unbundling and privatization of the nigerian electricity sector: reality or myth? University of Dundee, Scotland, United Kingdom. www.dundee.ac.uk/cepmlp/gateway/files.php?. Accessed 6 Aug 2015

Mandelli S, Barbieri J, Mattarolo L, Colombo E (2014) Sustainable energy in Africa: a comprehensive data and policies review. Renew Sustain Energy Rev 37:656–686

Marcellus IO (2009) Development planning in Nigeria: reflections on the national economic empowerment and development strategy (NEEDS) 2003–2007. J. Soc Sci 20(3):197–210

Mnse AO, Ojo AO (2009) Harvesting wind energy to solve Nigeria's energy crisis. J Eng Appl Sci 4(3):197–204

Mshandete AM, Parawira W (2009) Biogas technology research in selected sub-Saharan African countries—a review. Afr J Biotechnol 8(2)

Muhammad U (2012) Rural solar electrification in Nigeria: renewable energy potentials and distribution for rural development. SOLAR2012_0332

Nadabo SL (2010) Renewable energy as a solution to Nigerian energy crisis. Business Economics and Tourism, University of Applied Science

Naibbi AI, Healey RG (2013) Northern Nigeria's dependence on fuelwood: insights from nationwide cooking fuel distribution data. Int J Hum Soc Sci 3(17):160–173

National Bureau of Statistics (NBS) (2007) National account. Federal Republic of Nigeria, Abuja

National Energy Policy (NEP) (2003) Energy Commission of Nigeria (ECN). Federal Republic of Nigeria, Abuja. www.energy.gov.ng. Accessed 6 Aug 2015

National Energy Policy (NEP) (2006) Energy Commission of Nigeria (ECN). Federal Republic of Nigeria, Abuja. www.energy.gov.ng. Accessed 6 Aug 2015

National Energy Policy; Draft Revised Edition (NEP) (2013) Energy commission of Nigeria (ECN). Federal Republic of Nigeria, Abuja. www.energy.gov.ng. Accessed 6 Aug 2015

National Mirror (2014) Electricity: still in the dark hour. www.nationalmirroronline.net/new/electricity-still-in-the-dark-hour. Accessed 29 Dec 2014

National Planning Commission (NPC) (2009) Report on Vision 2020 National Technical Working Group on Energy Sector, Abuja, Nigeria

National Planning Commission, National Economic Empowerment, & Development Strategy (NEEDS) (2004) National economic empowerment and development strategy NEEDS, vol 2. International Monetary Fund

National Power Training institute of Nigeria (NAPTIN) (2015) www.naptin.org.ng. Accessed 6 Aug 2015

National Renewable Energy and Energy Efficiency Policy (NREEEP) (2014) Energy Commission of Nigeria (ECN) and Federal Ministry of Science and Technology (FMST) www.energy.gov.ng. Accessed 6 Aug 2015

Nee Nigeria (2015) Nigeria wind energy. www.neenigeria.com/Nigeria_wind_NEW.png. Accessed 6 Jan 2015
Ngala GM, Alkali B, Aji MA (2007) Viability of wind energy as a power generation source in Maiduguri, Borno state, Nigeria. Renew Energy 32(13):2242–2246
Nigeria Bureau of Public Enterprise (NBPE) (2015) Overview of the power sector. www.bpeng.org. Accessed 11 Aug 2015
Nigeria Customs Service (NCS) (2015) Automobile Importation in Nigeria from 1996 to 2012. www.customs.gov.ng. Accessed 6 March 2015
Nigerian Bulk Electricity Trading Plc. (NBET) (2015) www.nbet.com.ng. Accessed 6 Aug 2015
Nigerian Electricity Regulatory Commission (NERC) (2015) www.nercng.org. Accessed 6 Aug 2015
Nigerian National Petroleum Corporation (2015b) History of the Nigerian petroleum industry. Nigerian National Petroleum Corporation. www.nnpcgroup.com/NNPCBusiness/BusinessInformation/OilGasinNigeria/IndustryHistory.aspx. Accessed 8 Jan 2015
Nigerian National Petroleum Corporation (NNPC) (2015a) Refineries and petrochemicals. www.nnpcgroup.com/nnpcbusiness/midstreamventures/refineriespetrochem-icals.aspx
Nigerian national Petroleum Corporation (NNPC) (2015c) www.nnpcgroup.com. Accessed 6 Aug 2015
Nwaobi GC, Nigeria WA, Territory FC (2005) The Nigerian coal corporation: an evaluation of production performance (1960–1987) (No. 0501002). EconWPA
Nwoko A (2015a) Nigeria power plants. Department of Geography, Institute of Hazard, Risk and Resilience, Durham University, England
Nwoko A (2015b) Map of Nigeria showing yearly average global solar rasiation. Department of Geography, Institute of Hazard, Risk and Resilience, Durham University, England
Odesola IF, Eneje S, Olugasa T (2013a) Coal development in Nigeria: prospects and challenges. Int J Eng Appl Sci 4(1):64–73
Odesola IF, Samuel E, Olugasa T (2013b) coal development in Nigeria: prospects and challenges. Int J Eng 4(1):8269
Official Gazette of the Nigerian Bio-fuel Policy and Incentives (NBPI) (2007) Federal Republic of Nigeria. www.sunbirdbioenergy.com/docs/Nigeria_E10_POLICY_GAZETTED.pdf. Accessed 6 Aug 2015
Oil Shale and Tar Sands Programmatic EIS (OSTSPEIS) (2014) What is tar sands. www.ostseis.an/gov/guide/tarsands. Accessed 30 Dec 2014
Oji JO, Idusuyi N, Aliu TO, Petinrin MO, Odejobi OA, Adetunji AR (2012) Utilization of solar energy for power generation in Nigeria. Int J Energy Eng 2(2):54–59
Ojosu JO (1990) The iso-radiation map for Nigeria. Solar Wind Technol 7(5):563–575
Okoro OI, Chikuni E, Govender P (2007) Prospect of wind energy in Nigeria. In: Proceedings of the international conference on domestic use of energy
Oladimeji YU, Abdulsalam Z, Ajao AM, Adepoju SA (2015) Determinant of rural household poverty nexus fuel consumption among Fisherfolks in Kwara State, Nigeria. J Sci Res Rep 7(3):185–194
Opeh R, Okezie U (2011) The significance of biogas plants in Nigeria's energy strategy. J Phys Sci Innov 3
Oyedepo SO (2012) Energy and sustainable development in Nigeria: the way forward. Energy Sustain Soc 2(1):1–17
Oyedepo SO (2014) Towards achieving energy for sustainable development in Nigeria. Renew Sustain Energy Rev 34:255–272
Power Holding Company of Nigeria PLC (PHCN) (2010) National Control Centre Osogbo Generation and Transmission Grid Operations. 2010 Annual Technical Report, Osogbo, Nigeria
Presidential Task Force on Power (PTFP) (2015) www.nigeriapowerreform.org. Accessed 6 Aug 2015
Renewable Electricity Action Program (REAP) Federal Ministry of Power and Steel (2006) Federal Republic of Nigeria. www.iceednigeria.org/backup/workspace/uploads/dec.-2006-2.pdf. Accessed 6 Aug 2015

Renewable Electricity Policy Guidelines (REPG) Federal Ministry of Power and Steel (2006) Federal Republic of Nigeria. www.iceednigeria.org/backup/workspace/uploads/dec.-2006.pdf. Accessed 6 Aug 2015
Renewable Energy master Plan (REMP) (2005) Energy Commission of Nigeria (ECN) and United Nations Development Programme (UNDP). www.spidersolutionsnigeria.com/wp-content/uploads/2014/08/Renewable-Energy-Master-Plan-2005.pdf. Accessed 6 Aug 2015
Renewable Energy master Plan (REMP) (2012) Energy Commission of Nigeria (ECN) and United Nations Development Programme (UNDP). www.energy.gov.ng. Accessed 6 Aug 2015
Roadmap for Power Sector Reform (A Customer-Driven Sector-Wide Plan to Achieve Stable Power Supply) (RPSR) (2010) The Presidency, Federal Republic of Nigeria. www.nigeriaelectricityprivatisation.com/wp-content/uploads/downloads/2011/03/Roadmap-for-Power-Sector-Reform-Full-Version.pdf. Accessed 6 Aug 2015
Roadmap for Power Sector Reform (Revision 1) (RPSR) (2013) The Presidency, Federal Republic of Nigeria. www.nigeriapowerreform.org/content/Roadmap%20for%20Power%20Sector%20Reform%20-%20Revision%201.pdf. Accessed 6 Aug 2015
Rural Electrification Agency, Abuja (REA) (2015) www.rea.gov.ng. Accessed 6 Aug 2015
Rural Electrification Strategy and Implementation Plan (RESIP) (2006) Power sector reform team. www.rea.gov.ng/sites/default/files/Final%20FGN%20Rural%20Electrification%20Strategy%20-%20econ%20ONE.pdf. Accessed 6 Aug 2015
Rural Electrification Strategy and Implementation Plan (RESIP) (2014) Power sector reform team. www.power.gov.ng/National%20Council%20on%20Power/Rural%20Electrification%20Committe%20Recommendation.pdf. Accessed 6 Aug 2015
Safty4sea (2014) The Nigerian water ways. www.safty4sea.com. Accessed 24 Aug 2014
Sambo AS (1981) Renewable energy technology in Nigeria: resource availability and potential for application to agriculture. J Agric Technol 3(1):1–4
Sambo AS (2005) Renewable energy for rural development: the Nigerian perspective. ISESCO Sci Technol Vis 1:12–22
Sambo, AS (2008) Matching electricity supply with demand in Nigeria. International Association for Energy Economics (Fourth Quarter), pp 32–38
Sambo AS (2009a) Strategic developments in renewable energy in Nigeria. Int Assoc Energy Econ:16
Sambo, AS (2009b) The place of renewable energy in the Nigerian energy sector. In: World future council workshop on renewable energy policies, energy commission of Nigeria, Addis Ababa, vol 10
Sambo AS, Iloeje OC, Ojosu OJ, Olayande S, Yusuf AO (2006) Nigeria's experience on the application of IAEA s energy models (MAED & WASP) for national energy planning
Shaaban M, Petinrin JO (2014) Renewable energy potentials in Nigeria: meeting rural energy needs. Renew Sustain Energy Rev 29:72–84
Shehu AI (2012) Solar energy development in Northern Nigeria. J Eng Energy Res 2(1)
Standards Organization of Nigeria (SON) (2015) www.son.gov.ng. Accessed 6 Aug 2015
Tallapragada PV (2009) Nigeria's electricity sector-electricity and gas pricing barriers. International Association for Energy Economics (First Quater), pp 29–34
The Economist (2007) Nigeria: reforming the oil industry. www.economist.com. Accessed 6 Aug 2015
The Encyclopedia of Earth (2015) Energy profile of Nigeria. http://www.eoearth.org/view/article/152513. Accessed 4 Jan 2015
The Federal Government of Nigeria (FGN) (2005) Electric Power Sector Reform Act (EPSRA). www.nercng.org/index.php/nerc-documents/func-startdown/35/. Accessed 6 July 2015
The Presidency (2013) Roadmap for power sector reform: Revision 1 (RPSR) Federal republic of Nigeria. www.nigeriapowerreform.org/content/Roadmap%20for%20Power%20Sector%20Reform%20-%20Revision%201.pdf. Accessed 30 Dec 2015

The World Bank (2014) Energy consumption per capita. www.worldbank.org. Accessed 29 Dec 2015
TOPFORGE (2015) Nigerian Railway Corporation. www.topforge.co.uk/photographs/-NRC.htm. Accessed 8 Jan 2015
Transmission Company of Nigeria [TCN] (2015) www.tcnorg.com. Accessed 6 Aug 2015
US Energy Information Administration (EIA) (2014a) Nigeria overview. www.eia.gov. Accessed 29 Dec 2014
US Energy Information Administration (EIA) (2015a) Nigeria's crude oil production. www.eia.gov. Accessed 1 Jan 2015
US Energy Information Administration (EIA) (2015b) Nigeria's crude oil and condensate exports, 2012. www.eia.gov. Accessed 1 Jan 2015
US Energy Information Administration (EIA) (2015c) Crude oil consumption in Nigeria. www.eia.gov. Accessed 3 Jan 2015
US Energy Information Administration (EIA) (2015d) Nigeria's natural gas. www.eia.gov. Accessed 3 Jan 2015
US Energy Information Administration (EIA) (2015e) Natural gas reserves in the world. www.eia.gov. Accessed 3 Jan 2015
US Energy Information Administration (EIA) (2015f) World top natural gas flaring countries. www.eia.gov. Accessed 3 Jan 2015
US Energy Information Administration (EIA) (2015g) Nigeria's LNG exports in 2012. www.eia.gov. Accessed 5 Jan 2015
Vincent EN, Yusuf SD (2014) Integrating renewable energy and smart grid technology into the Nigeria electricity grid system. Smart Grid Renew Energy
VOX (2015) Why oil prices keep falling and throwing the world into turmoil. www.vox.com. Accessed 2 Jan 2015
World Green Building Council (WGBC) (2015) http://www.worldgbc.org/worldgbc/become-member/members/. Accessed 6 Aug 2015
World Meters (2015) Nigeria's population forecast. www.worldmeters.com. Accessed 1 Jan 2015
World Wind Energy Association (WWEA) (2008) World wind energy report 2008. WWEA: Bonn. www.wwindea.org/home/images/worldwindenergyreport2008. Accessed 6 Jan 2015
Zaku SG, Kabir A, Tukur AA, Jimento IG (2013) Wood fuel consumption in Nigeria and the energy ladder: a review of fuel wood use in Kaduna State. J Pet Technol Altern Fuels 4(5):85–89

Chapter 3
Literature Review

Abstract This chapter is divided into two sections: the first presents an insight into the various energy forecasting models, their approach, and classification; the second section presents some scholarly contributions in the economics literature which applied the LEAP model on various sectors in different countries of the world. The review in the second section is limited to the application area, findings, results, and synthesis in relation to the present study.

Keywords Backcasting and forecasting models · Energy modeling · Top-down and bottom-up models

3.1 Review of Energy Forecasting Models

The increasing availability of computers from the 1970s upward and the rise in environmental awareness due to climate change and oil crises saw the introduction of energy models (Kempton 1997; MacCracken 2004; Weart 2008). The models initially built were mostly for industrialized countries, and as such, their application to the developing countries was based on the assumption in relation to the developed countrys' situation (Shukla 1995). Furthermore, it was assumed that the future developmental projections would be similar to the historical projections of developed countries, which in reality was not the same. Although Urban et al. (2007) argued that the pre-2007 energy models were biased towards the real situation in developing countries, subsequent energy models have been upgraded to suit the needs of developing countries.

Most energy forecasting models use economic activity (GDP per capita, representing living standards) as driving force for energy-related issues. When comparing economic activity internationally, the local currencies are expressed in commonly known currencies such as the US dollars (Van Ruijven et al. 2008). They are developed specific to a nation or utility depending on the prevailing economic and market conditions (Suganthi and Samuel 2012).

N.V. Emodi, *Energy Policies for Sustainable Development Strategies*,
Frontiers in African Business Research, DOI 10.1007/978-981-10-0974-7_3

Forecasting using energy models is done through either backcasting[1] or forecasting.[2]

Some energy forecasting models are available and can be categorized under broad headings as briefly explained below.

- Time series models: are the simplest of all energy models which use time series trend analysis for extrapolating the future energy requirement. Time series models have been used in various energy economic studies to investigate the interrelations between energy–economic–environment variables for some time periods which are usually 25 years and above (Hamilton 1994).
- Regression models: are employed in such a way to test theories that the current values of one or more independent time series affect the current value of another time series.[3] They are used to forecast coal, oil, gas and electricity requirements, GDP projections as well as electric load forecasting. Much of the studies applying regression models are well established in the literature.
- Econometric models: are tools used by economists to forecast future developments in an economy (Hymans 2002). They are statistical models which specify the relationship that is believed to exist between the various economic quantities pertaining to a particular economic phenomenon under study. They are usually derived from a deterministic economic model by allowing for uncertainty, or from an economic model which itself is stochastic (Wooldridge 2012). Studies[4] applying econometric models include those dealing with energy demand and supply under some scenarios, projections of the total energy demand as a function of previous year's energy demand, price of energy, real income, and heating day.
- Decomposition models: is a technique used in research methodology to decompose an aggregated indicator, using either energy use or CO_2 emission, into its drivers (Sun and Ang 2000). Various studies such as Ang (1995a, b, 1996) have applied the decomposition method to investigate the changes in industrial energy consumption. Other studies have investigated the changes in other sectors as available in energy economics literature.
- Unit root test and cointegration models: unit root tests are used to test whether a time series variable is non-stationary using an autoregressive model. Two known types of tests include the augmented Dickey–Fuller test and the Phillips–Perron test, and they use the existence of a unit root as the null hypothesis.

[1]Backcasting usually refers to the process of exploring the past events given the information known to date. It usually starts by defining a desirable future and then works backwards to identify policies and programs that will connect the future to the presents (Brandes and Brooks 2005).

[2]Forecasting is the process of exploring the future events that have not been observed or determined and is based on current trend analysis (Holmberg and Robert 2000).

[3]Regression time series differ from the time series analysis previously mentioned, which focuses on comparing values of a single time series or multiple dependent times series at different point in time (Imdadullah 2013).

[4]Some studies include; Suganthi and Jagadeesan (1992), Rao and Parikh (1996), Suganthi and Williams (2000), Iniyan, Suganthi and Samuel (2006), and Roming and Leimbach (2015).

The cointegration tests which are preceded by various unit root tests are used to test the cointegration or long run relationships between two variables.

- Input–output models: are used to assess how social and economic changes will affect energy requirements in a country. They specifically show how industries are linked together through supplying inputs for the output of an economy.[5]
- Computable general equilibrium models: are a standard tool of empirical analysis that use economic data to estimate how an economy might react to changes in policy, technology, climate change, subsidy, different tax, and quota (Wing 2004).
- Bottom-up models: this includes partial equilibrium models (e.g. POLES, WEM, PRIMES), optimization models (LEAP, MARKAL, TIMES, MESSAGE), simulation models (NEMS-RSDM, MURE, REEPS, LEAP, MAED, NIA).[6] These models project energy demand for appliances and equipment in the residential, commercial, and industrial sectors, as well as other technologies in the sectors analyzed (MacNeil 2013). Some of these models will be further described in the subsequent subsections.

3.1.1 Approach to Energy Modeling

Energy models have two types of approaches and they are the top-down and bottom-up approaches. These are the two basic approaches to examine the linkages between the economy and specific GHG-emitting sectors such as the energy system. Top-down models evaluate the system from aggregate economic variables and they come from the way the energy modelers apply macroeconomic theory and econometric techniques to historical data on energy consumption, prices, income to model the final demand for goods and services, and supply it to the main sector (Pachauri et al. 2014).

On the other hand, bottom-up models work at a disaggregated level, and thus require extensive databases of empirical data to be able to build up the desired energy model (Kavgic et al. 2010). However, the application of the bottom-up model results in a more inclusive assessment and analysis of the energy technologies, policies, fuel mix, energy mix, and the entire energy system (Lkhagva 2014a, b). Table 3.1 shows the comparison between the top-down and bottom-up models.

From Table 3.1, it can be observed that the bottom-up models present a better approach in analyzing an energy system provided the complex data for the modeling are available.

According to Andersen and Termansen (2013) "bottom-up models present a comparative strength in their ability to investigate the impacts of energy policy on the portfolio of technologies that make up the supply and demand components of

[5]www.sjsu.edu/faculty/watkins/inputoutput.htm.

[6]See Herbst et al. (2012) for more on these energy models.

Table 3.1 Comparison between top-down and bottom-up models

Top-down models	Bottom-up models
Use an economic approach	Use an engineering approach
Give pessimistic estimates on best performance	Give optimistic estimates on the best performance
Cannot explicitly represent technologies	Allow for detailed descriptions of technologies
Reflect the available technologies adopted by the market	Reflect the technical potential
The most efficient technologies are given by the production frontier (which is set by the market behavior)	The efficient technologies can lie beyond the economic production frontier suggested by the market behavior
Use aggregated data for predicting purposes	Use disaggregated data for exploring purposes
Are based on the observed market behavior	Are independent of the observed market value
Disregard the technically most efficient technologies available, thus underestimating the potential for efficiency improvement	Disregard the market thresholds (hidden costs and other constraints), thus overestimating the potential for efficiency improvement
Determine the energy demand through aggregated economic indices (GNP, price elasticity), but vary in addressing the energy supply	Represent the supply technologies in detail using disaggregated data, but vary in addressing the energy consumption
Endogenize behavioral relationships	Assess the costs of the technological options directly
Assume that there are no discontinuities in historical trends	Assume that the interactions between the energy sector and the other sectors are negligible

Source Van Beeck (1999)

the energy system, in order to identify low-cost opportunities or design technology-based taxes, subsides or standards. The comparative strength of the top-down model is its ability to assess the macroeconomic costs of a policy shock and its economy-wide feedbacks on prices, commodity and factor substitution, income and economic welfare". This leads to further clarifications on the bottom-up and top-down models as the next subsection describes the classification of these models.

3.1.2 Classification of Existing Bottom-Up and Top-Down Energy Models

In this section, the classification of ten popular bottom-up and top-down energy models is carried out in nine classification patterns which include (i) purposes of the energy models, (ii) model structure (internal and external assumptions), (iii) the analytical approach, (iv) underlying methodology, (v) mathematical approach, (vi) geographic coverage, (vii) sectorial coverage, (viii) time horizon, and (ix) data requirement. This is shown in Table 3.2. This study employed the LEAP model (which is a top-down/bottom-up model) in energy demand and supply

Table 3.2 Classification of popular bottom-up and top-down energy models

	Developers	Purposes		Assumptions	Top-down versus bottom-up	Methodology	Mathematical approach	Level	Sectorial coverage	The time horizon	Data requirements
		General	Specific								
EFOM-ENV	European Commission DDG-XII F/1, Belgium	Exploring	Energy supply, energy and environment policy analysis and planning, emission reduction	Endogenous analysis of generation expansion. Input needed: demand projections/ scenarios, supply costs, (environmental) constraints	Bottom-up	Optimization	Linear programming/ dynamic programming	National	Energy producing and consuming sectors	Medium to long term	Quantitative, monetary, disaggregated
ENERPLAN	Tokyo Energy Analysis Group, Japan	Forecasting or exploring (depending on mode)	Energy supply, energy demand, matching demand and supply	Depends on mode	Top-down	Econometrics and simulation	Not available	National	Energy sector	Short to medium	Quantitative
ENPEP	International Atomic Energy Agency (IAEA), Austria	Forecasting, exploring	Energy demand, supply, matching demand and supply, environmental impacts	Demand: high degree endogenization supply: detailed description of end-uses and (renewable) technologies	Hybrid. Top-down for demand analysis and bottom-up for supply	Macroeconomic for demand, economic equilibrium for total energy system	Not available	Local, national	Entire economy	Short (1–3 years), medium, long (max 50 years)	Quantitative, monetary, aggregated and disaggregated

(continued)

Table 3.2 (continued)

	Developers	Purposes		Assumptions	Top-down versus bottom-up	Methodology	Mathematical approach	Level	Sectorial coverage	The time horizon	Data requirements
		General	Specific								
LEAP	Stockholm environmental Institute Boston, USA	Exploring, forecasting	Demand, supply, environmental impacts. Energy policy analysis, environmental policy analysis, biomass- and land-use assessment, preinvestment project analysis, integrated energy planning, full fuel cycle analysis	Demand: rather high degree of endogenization and description of all sectors in economy supply: simple description of end-uses and supply technologies, including some renewable	Demand: top-down, supply: bottom-up	Demand: econometric or macro-economic. Supply: simulation	Not available	Local, national, regional, global	All sectors	Medium, long term	Quantitative, monetary, aggregated/ disaggregated
MARKAL	International Energy Agency (IEA)/ETSAP	Exploring	Energy supply with constraints. The objective includes target-oriented integrated energy analysis and planning through a least cost approach	Low degree of endogenization, focuses only on the energy sector, detailed description of end uses and (renewable) energy technologies possible	Bottom-up	Toolbox/ optimization	Linear programming, dynamic programming	Local, national	Energy sector only	Medium, long term	Quantitative, monetary, disaggregated

(continued)

Table 3.2 (continued)

	Developers	Purposes		Assumptions	Top-down versus bottom-up	Methodology	Mathematical approach	Level	Sectorial coverage	The time horizon	Data requirements
		General	Specific								
MARKAL-MACRO	Brookhaven National Laboratory, USA	Exploring	Demand, supply, environmental impacts. Economy-energy-environmental analysis and planning	Neo-classical growth model with nested substitution (CES) between capital/labor aggregate and energy	MACRO part is top-down, MARKAL part is bottom-up	Macro-economic for MACRO and partial equilibrium through optimization for matching demand and supply in MARKAL	Dynamic programming (non-linear)	Local, national	All sectors	Medium, long term	Qualitative, monetary, aggregated, disaggregated
MESAP	IER, University of Stuttgart, Germany	Exploring, forecasting	Modular package. Demand, supply, environmental through different modules	Depends on module	Top-down (demand) and bottom-up (supply)	Econometric (demand), simulation or linear programming (supply)	(Among others) linear programming, dynamic programming	Local, national	All sectors through PLANET/MADE	Medium, long term	Quantitative, monetary, aggregated, disaggregated
MESSAGE-III	International Institute for Applied System Analysis (IIASA), Austria	Exploring	Energy demand and supply, environmental impacts. Generation expansion planning, end-use analysis, environmental policy analysis, investment policy	Detailed description of energy end-uses and (renewable) energy technologies	Bottom-up	Optimization	Dynamic programming	Local, national	Energy sector	Short, medium, long term	Quantitative, monetary, disaggregated

(continued)

Table 3.2 (continued)

	Developers	Purposes		Assumptions	Top-down versus bottom-up	Methodology	Mathematical approach	Level	Sectorial coverage	The time horizon	Data requirements
		General	Specific								
MICRO-MELODIE	CEA, France	Exploring	Energy demand, supply, environment. Analysis of macro-economic energy and environment linkages	Multi-sectorial analysis with a description of conventional energy technologies only, in particular for the electricity sector	Top-down with a detailed description of the energy sector	Macro-economic based on price equilibrium	Not available	National	All sectors, with a detailed description of the energy sector	Medium, long term	Quantitative, monetary, aggregated, disaggregated
RETscreen	CEDRL/ Natural Resources Canada	Exploring	Energy supply. Specially designed for renewable energy technologies	Detailed description of supply technologies for generation expansion	Bottom-up	Spreadsheet/ toolbox	Not available	Local, national	Energy sector	Not available	Quantitative, monetary, disaggregated

Source Van Beeck (modified) (1999)

forecasting along with their GHG emission using a scenario base approach. The LEAP model and the reason for the selection of the model are presented in the next chapter.

3.2 Literature Review and Synthesis

3.2.1 Literature Review

The LEAP model has been applied to a large number of scholarly articles and has been used by various government bodies and agencies for energy-related studies. Most studies including Shin et al. (2005) analyzed the impact of expand landfill gas (LFG) method of electricity generation on the cost of electricity generated as well as amount of greenhouse gas emissions using LEAP model for Korea by considering various different situations vis-a-vis business-as-usual scenarios of existing power plants. It has been found that adopting expand landfill gas (LFG) complemented with further forms of energy consumption would help in dwindling global warming potential up to 75 % in comparison with spontaneous release of CH_4 gas. The result of the study suggests that expand landfill gas (LFG) method of electricity generation in Korea would prove to be a good alternative for CO_2 dislocation from medium-term perspective and would result in supplementary energy earnings. Bose (1996) simulated the passenger transport sector for the capital city of India namely New Delhi city for various scenarios including business-as-usual (BAU), improvement in vehicular speed, increased share of public transport buses, introduction of mass rapid transport system (MRTS), and a combination of other scenarios from a medium- and long-term perspective.

An environmental and economic assessment was performed by Song et al. (2007) for Korea using the LEAP model. The study was based on energy policy changes particularly for climate change agreement and enhancement of CO_2 lessening expertise, scenario investigation via operating data for the CO_2 chemical absorption pilot plant which was installed in Seoul coal steam power plant (capacity of 2 Ton/day). The chief focus was on the impact of chemical absorption progression on energy and environmental scheme. The findings of the study show that CO_2 separation pilot plant in coal steam power plants had significance and there was inevitability of further investment and technology development for CO_2 mitigation technology in Korea. Dementjeva and Siirde (2010) modeled scenarios for energy sector of Estonia by extenuating the environmental impacts of electricity generation and by means of a new, lesser amount of environment-destructive technology using the Long-range Energy Alternatives Planning system (LEAP) framework. The result of the study gives insights to policymakers about energy modeling results from the perspective of Estonia wishing to adhere to the European Union's severe technological and environmental requirements.

The study by Huang et al. (2011) elucidated the importance of long-term forecasting of supply and demand of energy from Taiwan's perspective highlighting the fact of lack of natural resources in the country, profound reliance on energy imports by Taiwan and the of late stride of the country towards sustainable development. The study provided a glimpse of energy requirement and supply scenario in Taiwan and energy policies adopted over a period of time. The key objective of the study is application of the LEAP model for Taiwan's energy sector with an objective of comparing future energy requirement and supply trends including greenhouse gas emissions by considering alternative scenarios of energy policy and its impact on the evolving energy segment of Taiwan. The authors present outcome for different scenarios of implementation of energy policies, considering scenarios such as "business-as-usual" and aggressive energy-efficiency improvement policies. The authors also considered the case of on-schedule retirement of existing nuclear plants (three in number) and compared its impact with assumptions of lower economic growth on Taiwan's energy sector. The outcome of the study offers key insights for policy makers relating to adoption and implementation of future energy and climate policies in case of Taiwan.

In Taoa et al. (2011) study, they predicted scenarios which may act as an imperative indication for Chinese government to work towards having a low-carbon economy by considering four critical factors namely: Per capita GDP of China, energy structure in China, energy consumption in China and CO_2 emissions for measuring and quantifying the level of low-carbon economic development of China. By adopting the LEAP model, the authors successfully modeled for: (a) base scenario; (b) low-carbon scenario; and (c) frustrated low-carbon scenario in order to suggest China's low-carbon economic development (possible) levels for 2050. The results of the study elucidate that for China, the total terminal energy requirement (of coal) could be 6.095 billion tons (base scenario), 5.236 billion tons (low-carbon scenario) and 6.239 billion tons in 2050 (for frustrated low-carbon scenario). The study furthermore suggests that the main contributor for decrease in CO_2 emissions in China could be attributed to improved energy concentration. Auxiliary infiltration of renewable energy and fuel switching seems to have contributed to decrease in CO_2 emissions and concludes that these improvements would contribute significantly to achieve the goal of low-carbon economy ably coupled with and complemented by clean coal technology and energy efficiency in China.

The possible potential trends of energy demand, energy structure, and carbon emission were analyzed considering Beijing as a case study using LEAP model. The authors (Feng and Zhang 2012) used 2007 as the base year up to 2030 for three different scenarios namely: (a) business-as-usual (BAU); (b) basic-policy (BP) and (c) low-carbon (LC) scenario. The results of the study show that energy demand for Beijing under low-carbon (LC) scenario would be 88.61 million Mtce in 2030, which would be 55.82 % lesser than BAU scenario and 32.72 % lesser than BP scenario. Total carbon emissions for Beijing under low-carbon (LC) scenario in 2030 would be 62.22 % lesser than BAU scenario and 36.75 % lesser than BP scenario. The results suggest that the industrial sector in 2030 will account for the largest proportion for BP and LC scenarios compared to BAU scenario, and it has been found that building and transportation sector would play a major role in Beijing for effective control of energy consumption and carbon emissions for decades to come.

Household energy requirements for most of the developing countries are majorly sufficed from non-commercial fuels (example: animal dung, wood, crop residues) which have often resulted in human health deterioration as well as environmental hazards. In his study, Ibitoye (2013) focused on energy requirements for households in Nigeria under the framework that Nigeria would have access to clean affordable energy which is under the United Nation's (UN's) Millennium Development Goals (MDGs) by estimating energy requirements between 2005 to 2020 and having a base/reference period. For UN's MDGs to be achieved, the author estimated that household electricity utilization in Nigeria would increase by 41 % and the use of contemporary fuels would be more than double. This exodus to the use of contemporary fuels for cooking would result in diminution of overall fuel-wood consumption. Senshaw (2014) developed alternative scenarios (business-as-usual (BAU) and alternative scenario) for energy requirement of Ethiopia up to 2050 using the LEAP model. The results of the study provided crucial insights for Ethiopian policy makers while framing policies towards sustainable energy and development for decades to come.

A study similar to Bose (1996) was Bitos and Kiartzis (2014) who analyzed the energy demands and pollutant emissions for a variety of potential scenarios and policies using the LEAP software for Greek road transport sector considering 2010 as base year and extrapolating till 2035 by making estimations for: (a) business-as-usual (BAU) scenario; (b) substituting conventional fuels with alternate fuels leading to better fuel economy; (c) adoption of alternate technologies; (d) better efficiency in order to make possible predictions from policy perspective. The results of the study suggest that execution and adoption of enhanced fuel economy vehicles complemented by usage of alternative fuels and technologies would play a major role in reducing energy demand and alleviating pollutant emissions in the Greek transportation sector for decades to come.

The study of Mahumane and Mulder (2015) estimated aggregate trends in energy supply and demand due to projected gush in natural resources exploration in Mozambique, the energy infrastructure, and economic growth using the LEAP framework and using elasticities of GDP with respect to biomass consumption, structure of the sector, and vehicle ownership. The authors developed various scenarios and found that by 2030 primary energy production in Mozambique is expected to increase six times and majority of this would contribute to energy exports. With Mozambique all set to become a major energy player in decades to come, it has to strike a judicious balance between local and international export needs.

3.2.2 *Synthesis of the Literature*

From the literature review in the last subsection, a synthesis is carried out to bring out the essence of the previous studies in terms of application of Long-Range Energy Alternatives Planning System (LEAP) framework and mitigation of GHG emissions under various scenarios for different countries worldwide and in the process highlight the importance of this study from Nigeria's Long-term Energy Policy perspective.

Several studies in literature have focused on mitigation of GHG emission by shifting the energy utilization to demand side, while maximizing the potentials of diversification of the energy supply mix (Ghanadan and Koomey 2005 for California; Lin et al. 2010; Cai et al. 2008; He and Chen 2013 for China; Papagiannis et al. 2008 for Greece; Giatrakos et al. 2009 for Crete; Davoudpour and Ahadi 2006 for Iran; Phdungsilp 2010 for Thailand (Bangkok); Kadian et al. 2007 for India; Fadel et al. 2001 for Lebanon; Mondal et al. 2010 for Bangladesh; Mustonen 2010 for Lao). Other Studies have focused on GHG mitigation by shifting the energy carter to supply side (Zhang et al. 2010 for China; Islas et al. 2007 for Mexico; Takase and Suzuki 2011 for Japan; Kalashnikov et al. 2011 for Russia; Huang and Lee 2009 for Taiwan; Shin et al. 2005; and Lee et al. 2008 for Korea; Flores et al. 2011 for Central America—Honduras; Fadel et al. 2003 for Lebanon; Mulugetta et al. 2007 for Thailand).

This study falls into the category where it intends to lessen GHG emissions by imposing policies in both supply and demand side for Nigeria energy system, using the Long-Range Energy Alternatives Planning System (LEAP) framework and simulating scenarios (Similar studies include Dhakal 2003 for Nepal; Shabbir and Ahmad 2010 for Pakistan; Cai et al. 2008; Wang et al. 2010 for China; Bose 1996, 1998 for India). Based on the findings and syntheses of the literature, the future energy demand and supply in Nigeria is modeled using the LEAP model, as well as GHG emission.

In this study, emphasis is placed on (a) primary energy supply mix (i.e. coal, hydro, natural gas, crude oil, biomass, and other renewable resources), (b) transformation in the form of biofuel, charcoal, compressed natural gas, diesel, electricity, firewood, kerosene, liquefied petroleum gas, and petrol, (c) final energy consumption from the perspective of household, transport, industry, commercial/service as well as agriculture sector, (d) GHG emission reduction potentials in Nigeria, (e) least cost energy technology for Nigeria. Based on the synthesis of previous findings and studies, this work models and develops four scenarios which include: (i) reference scenario; (ii) low-carbon moderate scenario; (iii) low-carbon advance; and (iv) green optimistic scenario.

References

Andersen KS, Termansen LB (2013) Bottom-up and top-down modelling approach. IntERACT Model Working Paper No. 04. www.ens.dk/sites/ens.dk/files/info/facts-figures/scenarios-analyses-models/models/IntERACT/wp04_-_bottom-up_and_top-down_modelling_approach.pdf. Accessed 20 July 2015

Ang BW (1995a) Decomposition methodology in industrial energy demand analysis. Energy 20(11):1081–1095

Ang BW (1995b) Multilevel decomposition of industrial energy consumption. Energy Econ 17(1):39–51

Ang BW, Lee PW (1996) Decomposition of industrial energy consumption: tnt approach. Energy Econ 18(1):129–143
Bitos C, Kiartzis S (2014) Energy demand analysis and energy saving potentials in the Greek road transport sector
Bose RK (1996) Energy demand and environmental implications in urban transport case of Delhi. Atmos Environ 30(3):403–412
Bose RK (1998) Automotive energy use and emissions control: a simulation model to analyze transport strategies for Indian metropolises. Energy Policy 26(13):1001–1016
Brandes OM, Brooks DB (2005) The soft path for water in a nutshell. Friends of the Earth Canada
Cai W, Wang C, Chen J, Wang K, Zhang Y, Lu X (2008) Comparison of CO_2 emission scenarios and mitigation opportunities in China's five sectors in 2020. Energy Policy 36(3):1181–1194
Davoudpour H, Ahadi MS (2006) The potential for greenhouse gases mitigation in household sector of Iran: cases of price reform/efficiency improvement and scenario for2000–2010. Energy Policy 34(1):40–49
Dementjeva N, Siirde A (2010) Analysis of current Estonian energy situation and adaptability of leap model for Estonian energy sector. Energetika 56(1):75–84
Dhakal S (2003) Implications of transportation policies on energy and environment in Kathmandu Valley, Nepal. Energy Policy 31(14):1493–1507
Fadel M El, Zeinati M, Ghaddar N, Mezher T (2001) Uncertainty in estimating and mitigating industrial related GHG emissions. Energy Policy 29(12):1031–1043
Fadel M El, Chedid R, Zeinati M, Hmaidan W (2003) Mitigating energy-related GHG emissions through renewable energy. Renew Energy 28(8):1257–1276
Feng YY, Zhang LX (2012) Scenario analysis of urban energy saving and carbon abatement policies: a case study of Beijing city, China. Procedia Environ Sci 13:632–644
Flores WC, Osvaldo AO, Flores MA, Rivas FR (2011) Sustainable energy policy in Honduras: diagnosis and challenges. Energy Policy 39(2):551–562
Ghanadan R, Koomey JG (2005) Using energy scenarios to explore alternative energy pathways in California. Energy Policy 33(9):1117–1142
Giatrakos GP, Tsoutsos TD, Zografakis N (2009) Sustainable power planning for the island of Crete. Energy Policy 37(4):1222–1238
Hamilton JD (1994) Time series analysis, vol 2. Princeton University Press, Princeton
He LY, Chen Y (2013) Thou shalt drive electric and hybrid vehicles: scenario analysis on energy saving and emission mitigation for road transportation sector in China. Transp Policy 25:30–40
Herbst A, Toro F, Reitze F, Jochem E (2012) Introduction to energy systems modelling. Swiss J Econ Stat 148(2):111–135
Holmberg J, Robèrt KH (2000) Backcasting—a framework for strategic planning. Int J Sustain Dev World Ecol 7(4):291–308
Huang WM, Lee GWM (2009) GHG legislation: lessons from Taiwan. Energy Policy 37(7):2696–2707
Huang Y, Bor YJ, Peng CY (2011) The long-term forecast of Taiwan's energy supply and demand: LEAP model application. Energy Policy 39(11):6790–6803
Hymans SH (2002) Forecasting and econometric models
Ibitoye FI (2013) The millennium development goals and household energy requirements in Nigeria. SpringerPlus 2(1):1–9
Imdadullah M (2013) Time series analysis and forecasting. Basic statistics and data analysis. Lecture notes. www.itfeature.com/time-series-analysis-and-forecasting/time-series-analysis-forecasting
Iniyan S, Suganthi L, Samuel AA (2006) Energy models for commercial energy prediction and substitution of renewable energy sources. Energy Policy 34(17):2640–2653
Islas J, Manzini F, Masera O (2007) A prospective study of bio energy use in Mexico. Energy 32(12):2306–2320
Kadian R, Dahiya RP, Garg HP (2007) Energy-related emissions and mitigation opportunities from the household sector in Delhi. Energy Policy 35(12):6195–6211

Kalashnikov V, Gulidov R, Ognev A (2011) Energy sector of the Russian far East: current status and scenarios for the future. Energy Policy 39(11):6760–6780

Kavgic M, Mavrogianni A, Mumovic D, Summerfield A, Stevanovic Z, Djurovic-Petrovic M (2010) A review of bottom-up building stock models for energy consumption in the residential sector. Build Environ 45(7):1683–1697

Kempton W (1997) How the public views climate change. Environ: Sci Policy Sustain Dev 39(9):12–21

Lee S, Park JW, Song HJ, Maken S, Filburn T (2008) Implication of CO_2 capture technologies options in electricity generation in Korea. Energy Policy 36(1):326–334

Lin J, Cao B, Cui S, Wang W, Bai X (2010) Evaluating the effectiveness of urban energy conservation and GHG mitigation measures: the case of Xiamen city, China. Energy Policy 38(9):5123–5132

Lkhagva J (2014a) The long-term forecasting of the Mongolian energy demand and supply using the long-range energy alternatives planning system (LEAP) model (Master Thesis). Available from Department of Technology Management Economics and Policy Program, College of Engineering, Seoul National University, Republic of Korea

Lkhagva J (2014b) The long-term forecasting of the mongolian energy demand and supply using the long-range energy alternatives planning system (LEAP). Mater Degree Thesis, Seoul National University, Seoul, South Korea

MacCracken MC (2004) The discovery of global warming. Eos Trans Am Geophys Union 85(28):270–270

Mahumane G, Mulder P (2015) Introducing MOZLEAP: an integrated long-run scenario model of the emerging energy sector of Mozambique

McNeil MA (2013) Bottom-up energy analysis system-methodology and results

Mondal MdAH, Boie W, Denich M (2010) Future demand scenarios of Bangladesh power sector. Energy Policy 38(11):7416–7426

Mulugetta Y, Mantajit N, Jackson T (2007) Power sector scenarios for Thailand: an exploratory analysis 2002–2022. Energy Policy 35(6):3256–3269

Mustonen SM (2010) Rural energy survey and scenario analysis of village energy consumption: a case study in Lao People's Democratic Republic. Energy Policy 38(2):1040–1048

Pachauri RK, Allen MR, Barros VR, Broome J, Cramer W, Christ R, van Vuuren D (2014) Climate change 2014: synthesis report. Contribution of working groups I, II and III to the fifth assessment report of the intergovernmental panel on climate change. www.ipcc.ch/ipccreports/tar/wg3/index.php?idp=310. Accessed 19 July 2015

Papagiannis G, Dagoumas A, Lettas N, Dokopoulos P (2008) Economic and environmental impacts from the implementation of an intelligent demand side management system at the European level. Energy Policy 36(1):163–180

Phdungsilp A (2010) Integrated energy and carbon modeling with a decision support system: policy scenarios for low-carbon city development in Bangkok. Energy Policy 38(9):4808–4817

Rao RD, Parikh JK (1996) Forecast and analysis of demand for petroleum products in India. Energy policy 24(6):583–592

Roming N, Leimbach M (2015) Econometric forecasting of final energy demand using in-sample and out-of-sample model selection criteria

Senshaw DA (2014) Modeling and analysis of long-term energy scenarios for sustainable strategies of Ethiopia (Doctoral dissertation, Zentrale Hochschulbibliothek Flensburg)

Shabbir R, Ahmad SS (2010) Monitoring urban transport air pollution and energy demand in Rawalpindi and Islamabad using leap model. Energy 35(5):2323–2332

Shin HC, Park JW, Kim HS, Shin ES (2005) Environmental and economic assessment of landfill gas electricity generation in Korea using LEAP model. Energy Policy 33(10):1261–1270

Shukla PR (1995) Greenhouse gas models and abatement costs for developing nations: a critical assessment. Energy Policy 23(8):677–687

Song H-J, Lee S, Maken S, Ahn S-W, Park J-W, Min B, Koh W (2007) Environmental and economic assessment of the chemical absorption process in Korea using the LEAP model. Energy Policy 35:5109–5116

Suganthi L, Jagadeesan TR (1992) A modified model for prediction of India's future energy requirement. Int J Energy Environ 3(4):371–386

Suganthi L, Samuel AA (2012) Energy models for demand forecasting—a review. Renew Sustain Energy Rev 16(2):1223–1240

Suganthi L, Williams A (2000) Renewable energy in India—a modelling study for 2020–2021. Energy policy 28(15):1095–1109

Sun JW, Ang BW (2000) Some properties of an exact energy decomposition model. Energy 25(12):1177–1188

Takase K, Suzuki T (2011) The Japanese energy sector: current situation, and future paths. Energy Policy 39(11):6731–6744

Taoa Z, Liu Z, Zhao Changxin Z (2011) Research on the prospects of low-carbon economic development in China based on LEAP model. Energy Procedia 5:695–699

Urban FRMJ, Benders RMJ, Moll HC (2007) Modelling energy systems for developing countries. Energy Policy 35(6):3473–3482

Van Beeck N (1999) Classification of energy models. Tilburg University, Faculty of Economics and Business Administration

Van Ruijven B, Urban F, Benders RM, Moll HC, Van Der Sluijs JP, De Vries B, Van Vuuren DP (2008) Modeling energy and development: an evaluation of models and concepts. World Dev 36(12):2801–2821

Wang JN, Cai BF, Liu LC, Cao D (2010) Research and practice of low carbon society in China. www.caep.org.cn/english/paper/Research-and-Practice-of-Low-Carbon-Society-in-China-20100827.pdf. Accessed 2 Sep 2015

Weart SR (2008) The discovery of global warming: revised and expanded edition. Harvard University Press

Wing IS (2004) Computable general equilibrium models and their use in economy-wide policy analysis. Technical Note, Joint Program on the Science and Policy of Global Change, MIT

Wooldridge J (2012) Introductory econometrics: a modern approach. Cengage Learn

Zhang Q, Tian T, Zheng Y, Zhang L (2010) Fuel consumption from vehicles of China until 2030 in energy scenarios. Energy Policy 38(11):6860–6867

Chapter 4
Methodology, Data, and Scenario Development

Abstract This chapter is divided into three sections; the first section presents model used in this study, the reason for the selection of the model, and the algorithm of the LEAP and OSeMOSYS model. The second section describes the LEAP and OSeMOSYS data requirement, explains the data collection process, parameters, variables, and presents the datasets used in the development of the Nigeria LEAP Model. The last subsection explains the scenario development techniques, processes, and extensively discusses each scenario developed.

Keywords The LEAP model · LEAP-OSeMOSYS model · Modeling parameters · Data requirement and collection · Scenario development

4.1 The Model

This research will apply the Long-range Energy Alternatives Planning system[1] (LEAP) which is a scenario-based energy analysis and climate change assessment modeling tool, developed by Stockholm Environment Institute (SEI LEAP 2008). It designs different scenarios of future energy demand and environmental impact based on how energy is consumed, converted, and produced in a given region or economy under a range of values for parameters such as population increase, economic development, technology utilization, and inflation (Cai et al. 2008). With a flexible data structure, LEAP is user friendly and rich in technical specifications and end-use details (SEI 2008). The model has been extensively used at the local, national, and global scales to project energy supply and demand, predict environmental impact of energy policies, and identify potential problems.

[1]**LEAP citation**: Heaps 2012. Long-range Energy Alternatives Planning (LEAP) system. [Software version 2015.0.4] Stockholm Environment Institute. Somerville, MA, USA. www.energycommunity.org.

N.V. Emodi, *Energy Policies for Sustainable Development Strategies*,
Frontiers in African Business Research, DOI 10.1007/978-981-10-0974-7_4

4.1.1 Reason for the Selection of the LEAP Model

The LEAP model has some advantages compared to other models mentioned in the last chapter (Chap. 3) and these are the basis for the selection of the LEAP model presented in this study. These advantages include:

- Work scope: the LEAP model is able to work its way up from energy extractions, processing, conversion, transmission, up to end-use consumption by demand devices, under a range of assumptions.
- Data characteristics: the LEAP uses a flexible data structure which can be a Top-Down or Bottom-Up approach depending on the data available, or even decoupling approach.
- Policy analysis: with LEAP, energy policy analysts can develop and evaluate alternative scenarios by comparing the energy requirement, social costs and benefits, and their environmental impacts.
- Technology and Environmental Data (TED): the LEAP model is integrated with TED databases which give users information regarding technical characteristics, cost, and environmental effects of energy technologies.
- Graphical interface: the LEAP's interface is user friendly, rich in technical specifications and end-use details.

4.1.2 The Algorithm of the LEAP Model

The LEAP uses a framework for calculating energy consumption, transformation (electricity generation, oil refinery, charcoal production, coal mining), and carbon emissions and these are presented as follows.

4.1.2.1 Energy Consumption

The total final energy consumption is calculated as follows[2]:

$$\mathrm{EC}_n = \sum_i \sum_j \mathrm{AL}_{n,j,i} \times \mathrm{EI}_{n,j,i} \tag{4.1}$$

where EC is the aggregate energy consumption of a given sector, AL is the activity level, EI is the energy intensity, n is the fuel type, i is the sector, and j is the device.

The net energy consumption for transformation is calculated as follows:

$$\mathrm{ET}_s = \sum_m \sum_t \mathrm{ETP}_{t,m} \times \left(\frac{1}{f_{t,m,s}} - 1\right) \tag{4.2}$$

[2]Feng and Zhang (2012).

where ET is the net energy consumption for transformation, ETP is the energy transformation product, f is the energy transformation efficiency, s is the type of primary energy, m is the equipment, and t is the type of secondary energy.

The transport stock turnover is calculated as follows:

$$\mathrm{EV}_n = \sum_c s \times \frac{m}{fe} \tag{4.3}$$

where EV is the energy consumption in the transport sector, s is the number of vehicles (stock), m is the vehicle distance, fe is the fuel economy, n is the fuel type, and c is the vehicle type.

4.1.2.2 Transformation

The transmission and distribution module calculations take the domestic fuel requirement faced by the module and map those corresponding to the output fuels directly to the module input fuels. The total domestic requirements are then decreased by the outputs of the module and increased by the inputs to the module.[3] For each process p:

$$\mathrm{INPUT}_P = \frac{\mathrm{OUTPUT}_P}{\mathrm{EFFICIENCY}_P} \tag{4.4}$$

For a Transmission and Distribution module:

$$\mathrm{EFFICIENCY}_P = 1 - \mathrm{LOSSES}_P \tag{4.5}$$

where INPUT is the fuel or feedstock, OUTPUT is the electricity generated or the refinery/production output, EFFICIENCY is the efficiency of the power plants or refinery plants.

4.1.2.3 Carbon Emission

The carbon emission from final energy consumption is calculated as follows[4]:

$$\mathrm{CEC} = \sum_i \sum_j \sum_n \mathrm{AL}_{n,j,i} \times \mathrm{EI}_{n,j,i} \times \mathrm{EF}_{n,j,i} \tag{4.6}$$

where CEC is the carbon emission due to energy consumption, AL is the activity level, EI is the energy intensity, and $E_{n,j,i}$ is the carbon emission factor from fuel type n for equipment j from sector i.

The carbon emission from energy transformation is calculated as follows:

---

[3]Lazarus et al. (1997).

[4]Feng and Zhang (2012).

$$\mathrm{CET} = \sum_{s}\sum_{m}\sum_{t} \mathrm{ETP}_{t,m} \times \frac{1}{f_{t,m,s}} \times \mathrm{EF}_{t,m,s} \tag{4.7}$$

where CET is the carbon emission, ETP is the energy transformation product, f is the energy transformation efficiency, and $\mathrm{EF}_{t,m,s}$ is the emission factor from one unit of primary fuel type s consumed for producing secondary fuel type t through equipment m.

4.1.2.4 Costs

The total cost of sector is calculated as follows[5];

$$C = \sum_{i}\sum_{j}\left\{\left[\sum_{n}\left(e_{n,j,k}ep_{n}\right) + \sum_{k}\left(m_{k,j,i}mp_{k}\right) + fc_{j,i}\right]p_{j,i}\right\} \tag{4.8}$$

where C is the total cost of sector including equipment fixed costs and variable costs for raw materials and fuels, ep_n is the unit price of fuel type n, $mk_{,j,i}$ is the demand for raw material k per unit of production used in equipment j within production process i, mp_k is the unit price of raw material k, and $fc_{j,i}$ is the fixed cost per unit production through equipment j (within production process i).

4.1.2.5 Cost-Benefit Analysis

The cost-benefit analysis in LEAP calculates the costs of each part of the energy system such as the capital and operating maintenance costs of purchasing and using the technologies in the demand and transformation systems; the costs of extracting primary resources and importing fuels and the benefits from exporting fuels. Additionally, one can also optionally broaden the scope of the cost-benefit calculations to examine environmental externalities, by assigning costs to the emission of pollutants and any other direct social and environmental impacts of the energy system. LEAP performs cost-benefit calculations from a societal perspective by counting up all of the costs in the energy system and then comparing the costs of any two scenarios. LEAP can include all of the following cost elements[6]:

[5]Webmeets (2015).

[6]Webmeets (2015).

- Demand costs (expressed as total costs, costs per activity, or costs of saving energy relative to some scenario);
- Transformation capital costs;
- Transformation fixed and variable operating and maintenance costs;
- Costs of indigenous resources;
- Costs of imported fuels;
- Benefits of exported fuels;
- Externality costs from emission of pollutants;
- Other miscellaneous user-defined costs such as the costs of administering an efficiency program.

When setting-up costing in LEAP, it is very important to draw a consistent boundary around the system, so that LEAP will not double count costs and benefits. For example, if one counts the cost of fuels used to generate electricity it should not also count the cost of the electricity in an overall cost-benefit calculation. Figure 4.1 shows the costs required at different stages in LEAP's cost-benefit analysis.

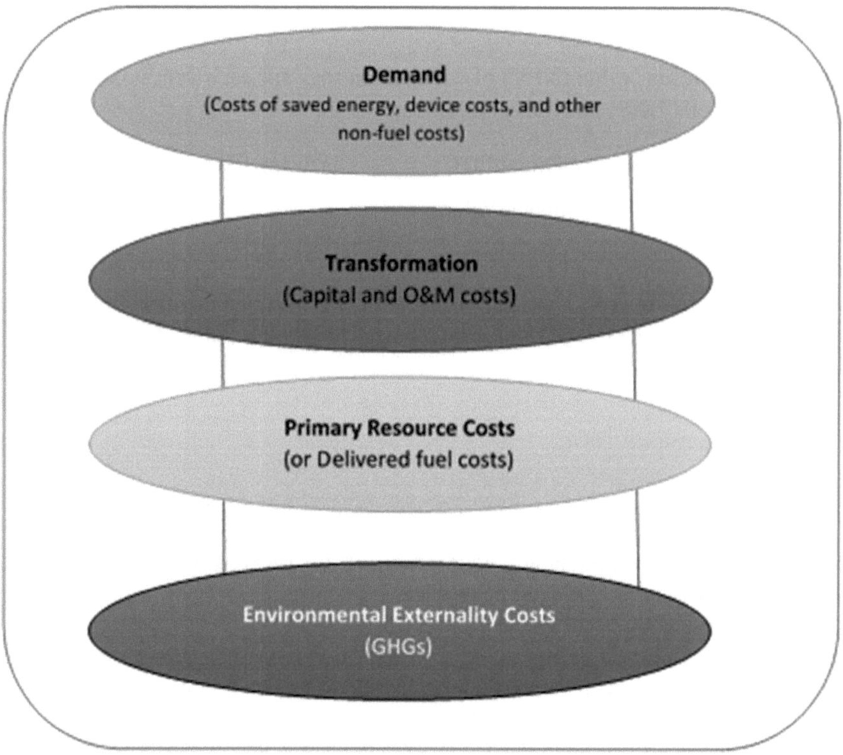

Fig. 4.1 Required costs of a cost-benefit analysis in LEAP model

4.1.2.6 Least-Cost Electricity Generation[7]

The least-cost electricity generation is calculated using the LEAP's optimization model called OSeMOSYS[8] to investigate a range of different technologies for generating electricity. The objective of this model is to estimate the lowest net present value (NPV) cost of an energy system to meet a given demand for energy or energy services. The optimization model is calculated as follows:

Least-Cost Electricity Generation Equations

Objective Function[9]

The total discounted cost of each technology over each year is summed and minimized to the following:

$$\sum_{y}\sum_{t}\text{TotalDiscCost}_{y,t} \tag{4.9}$$

where y represents year and t represents technology, while TotalDiscCost represents the total discounted cost as will be further described in Eq. (4.10).

Total Discounted Costs[10]

The total net present value (NPV) of each technology for each year is calculated as follows (Howells 2009; Howells et al. 2011):

$$\begin{aligned}\text{TotalDiscCost}_{y,t} &= \text{DiscOperatingCost}_{y,t} + \text{DiscCapitalInvestment}_{y,t}\\ &- \text{SalvageValue}_{y,t}\end{aligned} \tag{4.10}$$

where y represents year and t represents technology, while DiscOperatingCost represents the operating cost, DiscCapitalInvestment represents the discounted capital investment, and the SalvageValue represents the salvage value. Equation 4.10 is further described as follows.

[7]See Howells (2009) and Howells et al. (2011) for more detailed explanation on least-cost electricity generation using the OSeMOSYS functions.

[8]Open Source Energy Modeling System.

[9]The model is set up so that all costs incurred are attributed to a technology. Later, cost penalties will be introduced for under production (cost of energy not served) or over production of either energy or a commodity such as an emission.

[10]In order to derive the total costs incurred by each technology, the net present value operating, capital and salvage costs are summed by year for each technology. These values are stored in the variable: $\text{TotalDiscCost}_{y,t}$.

Operating Costs[11]

The total NPV operating cost for each technology for each year is calculated as follows (Howells 2009; Howells et al. 2011):

$$\text{DiscOperatingCost}_{y,t} = \text{OperatingCost}_{y,t} \times \text{DiscountFactor}_{y,t} \quad (4.11)$$

Thus,

$$\begin{aligned}\text{OperatingCost}_{y,t} = {} & \text{TotCapAnn}_{y,t} \times \text{FixedCost}_{y,t} \\ & \sum_{t} (\text{Activity}_{y,l,t} \times \text{VariableCost}_{y,t} \times \text{YearSplit}_{y,t})\end{aligned} \quad (4.12)$$

where y represents year and t represents technology, while l represents the time slice.[12] OpertingCost represents operating cost, DiscountFactor represents the discount factor.[13] From Eq. (4.12), TotCapAnn represents the total installed capacity of the power plant. FixedCost represents the fixed cost, VariableCost represents the variable cost, and YearSplit represents the year split (the fraction of a year).

Capital Costs[14]

The total NPV capital expenditure for each year and each technology is calculated as follows (Howells 2009; Howells et al. 2011):

$$\begin{aligned}\text{DiscCapitalInvestment}_{y,t} = {} & \text{CapitalCost}_{y,t} \times \text{NewCap}_{y,t} \\ & \times \text{DiscountFactor}_{y,t}\end{aligned} \quad (4.13)$$

where y represents year and t represents technology. CapitalCost represents capital cost, NewCap represents new capacity, and DiscountFactor represents discounted factor.

---

[11]Operating costs are calculated by summing the variable and fixed operating costs. The variable costs are a function of the technology's activity level during a time slice and the length of that time slice. These costs associated with each load region are summed to give an annual operating cost. The fixed cost is determined by the capacity of technology in the system. These costs are calculated for each technology, for each year. These values are stored in the variable: $\text{DiscOperatingCost}_{y,t}$.

[12]Annual demand is 'sliced' into representative fractions of the year, and this is necessary to assess times of the year when demand is high separately from times when demand is low, so that fuels that are expensive can be stored.

[13]The discount factor is the factor by which the future investment is multiplied in order to obtain the present value of the technology investment.

[14]The capital costs are calculated by multiplying capacity requirements by their respective capital costs. They are then discounted. These costs are recorded for each technology type as well as for the year in which the investment took place. These values are stored in the variable: DiscCapital $\text{Investment}_{y,t}$.

Salvage Value[15]

The value of each technology remaining at the end of the modeling period is calculated as follows (Howells 2009; Howells et al. 2011):

$$\begin{aligned}\text{SalvageValue}_{y,t} &= \text{CapitalCost}_{y,t} \times \text{NewCap}_{y,t} \\ &\quad \times \text{SalvageFactor}_{y,t}\end{aligned} \tag{4.14}$$

where y represents Year and t represents Technology. CapitalCost represents capital cost, NewCap represents new capacity, and SalvageFactor represents the salvage factor.

Fuel Production Constraints[16]

Fuel Production during each time slice and year must be greater than or equal to its use (consumption) by technologies plus any exogenous demand for that fuel, thus (Howells 2009; Howells et al. 2011)

$$\text{Production}_{y,l,f} \geq \text{Demand}_{y,l,f} + \text{Consumption}_{y,l,f} \tag{4.15}$$

where:

$$\text{Production}_{y,l,f} = \sum_t (\text{Activity}_{y,l,t} \times \text{OtptActvtyRatio}_{y,l,f} \times \text{YearSplit}_{y,l})$$

$$\text{and Consumption}_{y,l,f} = \sum_t (\text{Activity}_{y,l,t} \times \text{InptActvtyRatio}_{y,l,f} \times \text{YearSplit}_{y,t}) \tag{4.16}$$

[15]At the end of the modeling period, the stock of equipment which was invested in during the period has value which could be salvaged. This is dependent on (amongst other things) how it is depreciated, the life, and the time in which it is invested. (The NPV salvage value is calculated, and for simplicity is accounted for (by technology) in the year in which the new investment took place—rather than in the last year of the modeling period.) These values are stored in the variable: $\text{SalvageValue}_{y,t}$.

[16]There are several possible variations on this constraint which is the "main driver" of the model. For simplicity one is considered now. To start with, for each fuel an exogenous demand can be specified in terms of power consumed per time slice (e.g. In GJ/year or kW/year). This demand can be met from the activity of a technology. A technology's fuel production is related to its activity by an "output activity ratio". If a technology's activity, and output activity ratio for a given fuel is greater than zero, it produces that fuel. Similarly, the consumption or use of a fuel by a technology is related by its "input activity ratio". The model is constrained to produce more fuel than is used by energy technologies plus fuel that is exogenously demanded. The flexibility that comes with defining an input and output activity ratio allows the user to relate the capacities and activity in terms of the input of fuel (using an input activity ratio of 1, and an output activity ratio equal to the technology's efficiency (if only one fuel is produced). The capacity may be related to the technology's output (by having an output activity ratio of 1 and an input activity ratio equal to the reciprocal of the efficiency [if only one fuel is produced]). Note that how this is defined in turn determines how the costing parameters (capital and O&M) are quantified. Until "guided" by an interface, the user must be consistent. (The user may also define capacity in terms of kW, for example, but relate that to an activity in GJ/year by multiplying those ratios by 31.5, etc.).

where y represents year and t represents technology, while l represents the time slice and f represents the fuel. OtptActvtyRatio represents the output activity ratio, InptActvtyRatio represents the input activity ratio, while YearSplit represents the year split (the fraction of a year).

Capacity Constraints[17]

The capacity of each technology must be greater than its activity for each load region, thus (Howells 2009; Howells et al. 2011)

$$\text{Activity}_{y,l,t} \leq \text{TotalInstalledCapacity}_{y,t} \times \text{CapacityFactor}_{y,t} \tag{4.17}$$

where y represents year and t represents technology, while l represents the load region.

Activity Constraints[18]

The activity of each technology is limited by its annual availability, thus (Howells 2009; Howells et al. 2011);

$$\sum \text{Activity}_{y,l,t} \times \text{YearSplit}_{y,l} \leq \text{TotCapAnn}_{y,t} \times \text{AvailabilityFactor}_{y,t} \times \text{CapacityFactor}_{y,t} \tag{4.18}$$

where y represents year and t represents technology. YearSplit represents year split (the fraction of a year), TotCapAnn represents the total installed capacity of the power plant. AvailabilityFactor represents the availability factor which is $1 - M$ (M is the fraction of the year in during which planned maintenance takes place), CapacityFactor represents capacity factor.

[17]Each technology is able to accommodate an activity which uses, produces, or both uses and produces energy. This activity (currently measured in terms of power, such as kW/year or GJ/year) should be less than the de-rated capacity of that technology. The technology capacity is de-rated by a capacity factor. (If the capacity factor is kept at 1 there is no de-rating.) The total capacity of a technology is the sum of new investments made during the modeling period (which have not been retired) plus the any residual capacity left over from before the modeling period. In order to account for stations, invested in during the modeling period, which may be retired during the modeling period the notion of a station's "investment year" is introduced.

[18]In these constraints the activity of a technology is limited to a fraction of what its output would otherwise be—were it limited only by the capacity of the technology. Therefore, it may run at the maximum level (de-rated by the capacity factor) for some time slices, but for other time slices its output would be below this. (This serves to approximate planned outage in a facility such as a power station for example. The model determines the most economic time for this to happen.) This constraint is implemented by considering the activity rate for each load region and the length of the load region, the activity's annual availability as well as the total capacity of the technology in which the activity occurs. The total capacity determines the maximum potential activity during a load region (time slice).

Least-Cost Electricity Generation Model Sets, Variables, and Parameters

Model Sets

All model sets contain entries which are used as indexes to parameters and variables. The sets include the following (Howells 2009; Howells et al. 2011):

- Year: this set contains years over which the model is solved. Almost all variables and parameters are indexed by the model year for which it is solved.
- Investment year: this set is used to track the years in which investments are made. Investment years are identical in number to model years. It is separate from the model year set as it will be used to index investments by their "investment year" over the modeling period.
- Technology: this is the set of all technologies represented in the application.
- TimeSlice: this set contains the time slices per year. At this point, the same number of time slices is carried through for each year. The number of slices is also limitless.
- Fuels: this set contains the fuels used in the model. The model is constrained so that fuel production should be larger than the specified exogenous demand and technology fuel use.
- (OpFlows: this is to be added and will include all "flows" that the user will relate to operation—this could include emissions for example)
- (ConFlows: this is to be added and will include all "flows" that the user will relate to construction—this could include demands for materials for example).

Model Variables

The model variables used are presented as follows (Howells 2009; Howells et al. 2011):

- NewCap: new investment in capacity by the model year in which the investment was made, for each technology.
- TotCapAnn: this gives the total capacity available by technology for each year. It consists of the sum of the capacities of different vintages which are available for the given model year as well as any residual capacity from previous years.
- Activity: this variable represents the "activity" of the technology. Its unit is in terms of power—of energy out per year where the energy out is related by an output to activity ratio. (Conversely the energy used is related to the activity by an input to activity ratio) Activity is calculated for each technology, for each year, for each time slice. To convert activity from power to energy, it is multiplied by the length of the time slice over which it is active.
- Production: gives the total output of each fuel from all technologies for each time slice.
- Use: as with the production variable, this gives the total consumption—or use—of each fuel from each technology.
- CapitalInvestment: for each model year and each technology this variable is derived by multiplying new capacity investment (NewCap) with the capital cost of each technology.

- DiscCapitalInvestment: for each model year and each technology, this variable gives discounted capital investment.
- SalvageValue: this is the discounted salvage value calculated by the technologies remaining at the end of the model period.
- OperatingCost: this is the total annual operating cost for each technology. It includes both the variable and fixed costs.
- DiscOperatingCost: this is the discounted total annual operating cost for each technology.
- VarOpCost: the variable operating cost is calculated for each time slice, each technology, for each year.
- AnnVarOpCost: for each technology and year the variable operating cost is calculated. This is calculated by summing the variable operating costs incurred for each load region in each year.
- TotalDiscCost: for each technology and for each year, the NPV capital, operating, and salvage costs are summed in this variable.

Model Parameters

The model parameters presented below are those connected to the LEAP model used in this study, and they are as follows (Howells 2009; Howells et al. 2011):

- Capacity Factor: this is defined for each year and each technology. It is used to convert annual capacity to available capacity for each time slice.
- Availability Factor: this is defined for each year and each technology. It is used to simulate "planned outages" and indicates the maximum the technology may run for the whole year.
- Vintage Matrix: this is defined for each model year, for each technology vintage and each technology and is used to construct the total installed capacity variable. It is a function of the technology life, as well as when that technology was investment was made.
- Salvage Factor: this is the (discounted) salvage value of a technology at the end of the modeling period, represented as a fraction of the initial per unit capital cost. It is determined for each model year as well as each technology. It is calculated by estimating the depreciated value of the technology at end of the modeling period. (In the optimization, this is required to determine the full NPV costs of technology investment. At present, flexibility is left to the user as to how to calculate this.)
- Residual Capacity: this is defined for each model year and each technology. It is the capacity left over from a period prior to the modeling period.
- Output Activity Ratio: gives the output of fuel as a ratio to the activity of the technology. It is defined by year (as the technology could degrade), by technology, by time slice and for the fuel which is produced.
- Input Activity Ratio: gives the input (use) of fuel as a ratio to the activity of the technology. It is defined by year (as the technology could degrade), by technology, by time slice, and for the fuel which is used.
- Capital Cost: The capital cost of each technology is given as a function of the technology as well as the year in which the technology was invested. The cost

is the NPV cost for a new power station (presently calculations to determine the NPV cost as a function of the station's build profile are undertaken in the Excel mock-up interface.)

- Variable Cost: The variable cost is defined for each technology in each year and is the cost per unit of activity of that technology.
- Fixed Cost: The fixed cost, also a function of technology as well as the model year, is the cost per unit of capital stock (or installed capacity) of a particular technology.

4.2 The LEAP Data Requirement[19]

The LEAP as a general-purpose software tool typically used to build a wide variety of energy models may be difficult to definitively provide its data requirement (Heaps 2006). A key benefit of LEAP is its low initial data requirement[20] which depends on the type of energy modeling being carried out (i.e. bottom-up or top-down approach). However, the LEAP model requires comprehensive knowledge on data collection, understanding of the energy system, and time-consuming efforts especially in the data collection and input period. The type of data required for analysis on the LEAP model is described below and can be found in Heaps (2006).

4.2.1 Demographic Data

This is usually the general data of a country which include national population data, rates of urbanization, average household sizes, household growth rate, population growth rate, and urbanization growth rates. In some modeling, population by region, male/female population and age structure of population may be required. All these are entered into the "Key Assumption" of the data tree in the LEAP model.

4.2.2 Economic Data

The economic data include GDP/GNP data, value added by sector/subsector, average income levels, and interest rates. Other data include production of energy-intensive materials (output in tons or US$ per steel), transport needs (passenger-km, ton-km, vehicle-km), income distribution.

---

[19]Please note that the data requirement in this section is quite different from those required by the **Least-Cost Electricity Generation in Sect.** 4.1.2.4, because the data have already been described there.

[20]www.energycommunity.org/default.asp?action=47#LowDataRequirements.

4.2.3 General Energy Data

These are usually found in the National energy balances with data on energy consumption and production by sector or subsector in an economy. Most of these data are found in national statistical bodies or agencies or energy-related agencies as the country case may be. If the data are not available in the country, they may be available from the International Energy Agency (IEA) published energy statistics. Other data include national energy policies and plans, annual statistical reports with information on production, consumption, etc., of oil, natural gas, coal, charcoal, LPG, CNG, and other relevant fuel.

4.2.4 Demand Data

- Activity Levels: LEAP's demand analysis works by forecasting future energy consumption as the product of two factors: activity levels and energy intensities. Activity levels are simply a measure of the economic activity in a sector, and you can choose what data to use for this purpose. For example, in the household sector the user may choose to use the number of households as the activity level, in the cement industry you might use tons of cement production, and in the transport sectors you may choose to use ton-kms (for freight transport) and passenger-kms (for passenger transport). The user will need to collect data describing the current, historical, and future projections of whatever data the user chooses to use for his/her activity level variables. The user may need to consult national statistical reports or contact governmental or academic organizations working in specific sectors (industry, commerce, transport, households, etc.).
- Energy intensity data is often very difficult to come by. If the user is preparing an aggregate analysis he/she will likely be able to combine their activity level data with national energy consumption statistics and energy balances to calculate historical energy intensity values by sector and by fuel. In other words, for historical data, *energy intensity = total energy consumption/activity level.* For your forward looking scenarios you will instead use LEAP to calculate the total energy consumption by projecting the energy intensity and activity level. That is: *total energy consumption = energy intensity x activity level.*
- Other useful sources of energy demand data include recent social surveys or energy consumption surveys that analyze how energy is consumed in different sectors of the economy, and reports from utilities and private companies on sales of different energy forms (electricity, natural gas, oil products). If possible, the user should try to get data disaggregated by sector and by consumer category.
- If the user is creating a more detailed analysis, he/she will likely also need information on the stocks, technical characteristics (efficiency, specific fuel consumption), costs, and environmental loadings of major energy-consuming

devices in different sectors. For example, if the user wants to focus on road transport energy use, you would need data describing the stocks and sales of vehicles, their fuel economy, and some estimate of their average on-road life expectancy.

4.2.5 Transformation Data

In a general case, transformation analysis requires that the user prepare a complete picture of how energy is extracted, converted, and transported in the modeled energy system. This requires data on the flows of energy into and out of major processes, as well as information on the efficiency, costs (capital, operating and maintenance, and fuel costs) and environmental loadings associated with each major process.

- Electric sector: Generally, the user will need data describing the current and historical installed capacities (MW), efficiencies, costs (capital, operating and maintenance, and fuel costs) and actual dispatch (MW-HR) of the various types of electric generating plants in your country. The user will also need information on the seasonal load shape for his/her country's electric system and the maximum availability and dispatch priority of each different type of power plant. Capacity expansion plans, if existent, can be very useful for establishing forecasts of how the electric system is likely to evolve in the future. In addition to collecting data on generation, you should also collect data describing transmission and distribution losses including both technical and non-technical losses. The user may wish to analyze this sector separately from the dedicated electric generation sector. For this sector, the user's data should include the production efficiencies of both electricity and heat. In many countries, rural electrification is a key issue, so you may wish to collect relevant data describing rural electrification rates for different geographic regions.
- Oil Refining: If oil refining is an important sector in the model development, the user will need to collect data on the different products produced by your refineries, the efficiency, and the capacity of the refineries.
- Extraction sectors: If extraction sectors such as coal mining or oil and gas production are important, the user will need data describing the efficiency and capacity of these sectors as well as information on the fuels produced and the energy consumed during extraction.
- Renewables: Renewable energy is becoming increasingly important in many countries and may be an important focus of any GHG-mitigation analysis. Collect data describing the current installed capacities, efficiencies, costs, and expansion plans for any relevant renewables such as wind, geothermal, municipal solid waste, solar, etc.
- Biomass: If wood or other biomass fuels are important in the country, the user will have to collect whatever data are available on the consumption and

production of those fuels. Wood fuel (or firewood or fuel wood) surveys can be an important source of data for estimating the sustainability of production of wood fuels.
- Other Sectors: Other conversion sectors that may be important include charcoal making, ethanol production, and synthetic fuel production from coal.

4.2.6 Environmental Data

For a first cut GHG-mitigation assessment, the user may be able to rely on the basic "Tier 1" emission factors published by the IPCC (and included in LEAP). However, as the user refines his/her analysis, they may wish to collect local emission factors estimates that reflect the fuel and technology characteristics of devices used in the country. For example, cars in the country may have particular emissions characteristics. It is particularly important to have data on the chemical composition of the fuels used in the country as this can be used to refine the emission factor estimates from different devices. The IPCC's online EFDB database is a key source of data on emission factors.[21]

4.2.7 Fuel Data

LEAP includes a good default list of fuel and their characteristics (energy content, chemical composition) that should meet the needs of most studies. However, the user should be sure to adjust the energy, carbon, and sulfur contents in this list to reflect the characteristics of the fuels used in the country being analyzed. In particular, the characteristics of coal and biomass fuels vary greatly between (and even within) countries and uses. In addition to their physical characteristics, the user will also require data describing the production costs of any primary fuels produced in the country and the import and export costs of any relevant fuels.

4.2.8 Data Collection Process

The Nigerian LEAP model was developed to forecast the future energy demand, supply, and GHG emission under the Reference scenario and alternative scenarios (explained in later subsections). The data collected for the Nigerian LEAP model were from many sources which include the National Bureau of Statistics in Nigeria (NBS) (primary data source), Energy Commission of Nigeria (ECN) (for

[21]This is available at: http://www.ipcc-nggip.iges.or.jp/EFDB/main.php.

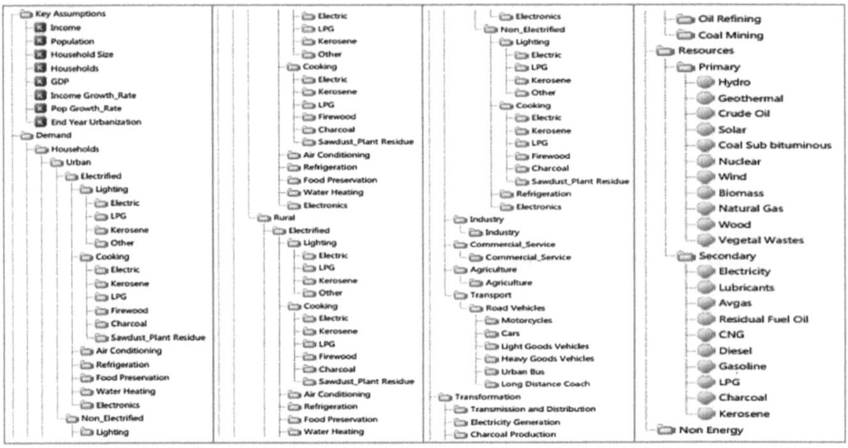

Fig. 4.2 The Nigerian LEAP model data branches

national policies, some primary data and plans), National Control Center, Osogbo in Nigeria (NCC) (power plants data on energy generated), the Petroleum Products Pricing Regulatory Agency in Nigeria (PPPRA), the World Bank, International Energy Agency (IEA), the United States Energy Information Administration (EIA), the International Association of Public Transport (UITP) and African Association of Public Transport (UATP), National Surveys by bodies and independent researchers, credible research articles, newspaper and website articles, and publications. The datasets used are presented in Table B.1–B.29.

The Nigerian LEAP dataset were vast but carefully divided into five subsectors of energy demand which include households, industry, transport, commercial/service, and agriculture. The household sector was split into two main branches and two sub-branches under the main branches: the urban household was divided into electrified and non-electrified, while the rural household was divided into rural electrified and non-electrified. This is shown in Fig. 4.2 which is the Nigerian LEAP model data branches.

In the industrial sector, energy use is divided into shares of energy use (i.e. gas, coal, electricity, biomass), and this is due to the data availability. There is a similar situation in the commercial/service and agriculture sectors. The transportation sector is divided into six branches: motorcycles, cars, light duty vehicles, heavy duty vehicles, urban bus, and long-distance coach. The transformation branch[22] is divided into five branches: transmission and distribution, electricity generation, charcoal production, oil refinery, and coal mining. The final resource branch is the branch which includes the primary and secondary energy resources.

[22]The transformation sector includes the transmission and distribution of electricity and natural gas network, electricity generation, charcoal production, oil refining and coal mining.

4.3 Scenario Development

The scenarios were developed using the Reference scenario (Ref), from which three other alternative scenarios were developed to provide a low carbon-green growth development in Nigeria. The alternative scenarios developed are Low Carbon Moderate (LCM), Low Carbon Advance (LCA) and Green Optimistic (GO) scenario. In the development of these scenarios, the Ref. scenario assumed that the energy policy goals of the Nigerian government through its policy documents (NEMP 2014; NEP 2013; NREEEP 2014), will be achieved by the end of 2040 from 2010.

Thus, the Nigeria LEAP model was developed from 2010 which is the base year, to 2040 which is the target/end year. The alternative scenarios (i.e. LCM, LCA and GO) were developed from the same base year and target year period (30 years), and inherited the characteristic features of the base year from the Ref. scenario. However, three different policy pathways were followed in the three alternative scenarios, of which their individual goal was to use a low carbon approach to match energy demand with supply at the least cost until 2040. The scenarios are outlined and discussed below, while the key assumption parameters for the Nigerian LEAP model is shown in Table 4.1 and a summary of the features of each scenario is presented in Table 4.2.

Table 4.1 Key assumption parameters for the Nigerian LEAP model under the Ref, LCM, LCA, and GO Scenarios

Key assumption	Ref**		LCM		LCA		GO	
	2010	2040	2010	2040	2010	2040	2010	2040
GDP (million USD)	369	1255*	369	1476*	369	1587*	369	2177*
GDP growth rate (%)	8		10		11		13	
Income (Thousand USD)	2310	5705	2310	5914	2310	6329	2310	6676
Income growth rate (%)	4.9		5.2		5.8		6.3	
Population (Million)	160	304	160	318	160	328	160	337
Population growth rate (%)	2.55	3	2.55	3.30	2.55	3.50	2.55	3.70
Households (Million)	37	72	37	80	37	84	37	87
Household size (people)	5							
Household growth rate (%)	3.16		3.9		4.2		4.5	
Urbanization rate (%)	44	70	44	75	44	80	44	73
Industry growth rate (%)	6	6.5	6	7.1	6	7.5	6	7.7
Commercial/Services growth rate (%)	12	12.5	12	13.2	12	13.6	12	13.9
Agriculture growth rate (%)	6	6.5	6	7.3	6	7.4	6	7.8

*Denotes Billion, **The figures in the Ref scenario are based on the NEMP (2014)

Table 4.2 Summary of the features in each scenario

	Ref	LCM	LCA	GO
Driving philosophy	Follows the Government's most-likely-developmental-pathways for the energy system	Driven by cheaper capital, cost, readily available fuels to improve power supply, and moderately reduce energy demand	Motivated by a cleaner fossil fuel power technologies and an aggressive reduction in energy demand	Based on a low-carbon-green growth economy in view of mitigating global climate change, reduce energy poverty, and ensure energy sustainability.
Scenario characteristic	• Current trend of energy consumption continues in all sectors • No supply side diversification of energy source (BAU case)	• Moderate improvement in energy efficiency in all sectors • CFL bulbs to replace 85 % incandescent bulbs • Increase in gas power plants, small share of renewables • Introduction biofuel in transport sector • Moderate reduction in electricity T&D and natural gas loss	• Aggressive improvement in energy efficiency in all sectors • LED bulbs to replace 70 % incandescent bulbs • Increased capacity of efficient low carbon fossil fuel power plants, increased share of renewables • Introduction of LPG fuel options in addition to biofuel to complement conventional fuels in the transport sector • Improved reduction in electrical and natural gas losses	• Provision of Energy efficiency and solar PV systems for the demand side • Phase-out incandescent bulbs and replace with CFL 60 % and LED 40 % • Reduction in the share of fossil fuel plant capacity and the increase share of renewables on the supply side • CNG, LPG and biofuel options with a reduced share of conventional fuels • Reduction in electrical and natural gas losses to the barest minimum
General assumptions	Increased electricity access to 100 % by 2030			

(continued)

Table 4.2 (continued)

	Ref	LCM	LCA	GO
Scenario-specific assumptions	Demand side: Business-as-usual case in all sectors	Demand side: CFL introduction, moderate energy efficiency in appliances	Demand side: LED introduction, aggressive energy efficiency in all sectors	Demand side: introduction of both CFL and LED with reduction in incandescent bulbs, advance approach towards energy efficiency in all sectors
	Supply side:	Supply side:	Supply side:	Supply side:
	Installed capacity by 2040 = 155283 MW	Installed capacity by 2040 = 170500 MW	Installed capacity by 2040 = 180500 MW	Installed capacity by 2040 = 181000 MW
	Estimated share of fuel type:	Estimated share of fuel type:	Estimated share of fuel type:	Estimated share of fuel type:
	Fossil fuel = 90 %	Fossil fuel = 80 %	Fossil fuel = 60 %	Fossil fuel = 30 %
	Renewables = 10 %	Renewables = 20 %	Renewables = 40 %	Renewables = 70 %

4.3.1 The Base Year

In the base year (2010), Nigeria had a gross domestic growth (GDP) of US$ 369 Million which has a growth rate of 6.3 %. Average income was US$ 2310 and the country's population was 160 million with a growth rate of 2.55 %. According to the National Bureau of Statistics (NBS 2015), there are 37 Million households in Nigeria with an average of 5 people per house and have a growth rate of 3.16 % each year. The country's urbanization rate was 44 % and access to electricity was 48 % in the base year. The percentage of urban population with access to electricity was 79.80 % while that in rural area with access to electricity was 34.90 %.

Energy consumption for cooking has been observed to have the highest share of intensities in the household sector in Nigeria (Ibitoye 2013; Ogwumike et al. 2014; Akinyele et al. 2014; Oteh et al. 2015). Although a large share of the households in both the rural and urban areas use kerosene for cooking, LPG and electricity have continued to play a crucial role in the choice of fuel for cooking. In the rural areas, firewood (or fuel wood) has a competitive share with kerosene, while LPG and electricity consumption is relatively small compared to other source of fuel for cooking. From the characteristics features of household cooking technologies shown in Table B.24, the number of stoves which use firewood and kerosene is high, as is their operating time. Other features of the household technologies, their percentage shares and intensities are shown in Table B.11–B.12.

Other sectors of the economy that were observed in the base year are industry, commercial/service, and agriculture sector. The value added for the industry sector was 80 million dollars with a growth rate of 6 %, while that of the commercial/service and agriculture were 196 million and 86 million respectively. The type of energy used in the sectors and their consumptions are shown in Table B.8. In the transport sector, six types of vehicles were used in this study and they include motorcycles, cars, light goods and heavy goods vehicles, urban bus, and long-distance coach. The fuel type, fleet, vehicle activity, and fuel efficiencies are shown in Table B.15–B.16.

In the power (electricity) sector, electricity generation is mainly from thermal and hydropower stations. The thermal power stations are made up of only natural gas power plants with majority being single cycle gas turbines (SCGT), and a small share of combined cycle gas turbine and gas-fired steam turbine (GFST). The average availability of the thermal power plants is observed to be half of the installed capacity, but the hydro power plants have a higher average availability as compared with the thermal power plants. On the part of transmission and distribution for the power sector, a 20 % loss rate was recorded in 2010. Details of the electricity generation are shown in Table B.27.

On the natural resources and refinery operations recorded in the base year, the conventional energy resources observed were crude oil, natural gas, and coal (sub-bituminous). The secondary energy supplies which are from the conventional energy resources produced in the refinery are gasoline, diesel, LPG, Kerosene, lubricant, avgas, CNG, residual fuel oil, charcoal, and firewood (from biomass). Details of the data used for refineries of crude oil, charcoal production and coal mining in the base year are shown in Table B.5–B.7, B.10, and B.17–B.19. From the data and characteristic features of the socio-economic activities in the base year (2010), the development of the scenarios were established following the Reference scenario which will be discussed in the following subsection.

It is very **important to note** that when developing an energy model using the LEAP, energy efficiency measures can be imputed in two ways:

- By including another variable called "**efficient**" alongside the **existing** technology. If the government (the user in this case) intends to increase energy efficiency of a particular technology such as refrigerator in the household sector in 2015, the share of efficient refrigerator variable will be increased from 2015. If the government intends to have 70 % share of efficient refrigerator by 2040 starting from 2015, the "**efficient refrigerator**" variable is interpolated into the technology from the reference or business-as-usual case. The results will then show, for example, Demand-Household-Refrigeration-existing and efficient. This same case will be done for all technologies in the sector, including transport, industry, etc.
- The other way of imputing energy efficiency measures is through the decrease in "**final energy intensity**" by end/target year. If the government intends to increase energy efficiency by a certain amount, the existing technology's final energy intensity is then **interpolated** to a given decrease by the target year. To **interpolate**, the following command is inserted into the technology variable's final energy intensity: Demand-Household-Refrigeration-Existing-Final Energy Intensity-Interp (2040, –10) or Interp[23] (2020, –2, 2030, –5, 2040, –10). It however depends on how the user wants the energy efficiency target to be for a particular technology in a sector. In some situations, the final energy intensity value can be imputed directly into the LEAP model. For example, if the base year value for household refrigerator's **final energy intensity** is 400 kw-h/household, the user can set a final energy intensity value for the particular technology to be 200 kw-h/household by 2040. To do this, the following command is imputed into the LEAP model: Demand-Household-Refrigeration-Final Energy Intensity-200 kw-h/Household. The results will then show, for example, Demand-Household-Refrigeration-efficient-value. Either of the two ways (i.e. the Interp or Final Energy Intensity) will still produce the same result, which is Demand-Household-Refrigeration-efficient-value.

The same way of imputing energy efficiency measures was also applied in all demand sectors in the Nigerian LEAP model.

[23]Interp means to intercept a given value in the model in a particular year or period of time or years.

4.3.2 The Reference Scenario (Ref)

The development of the Ref. scenario is in line with the concept of the most-likely-developmental-pathways of the government in the Nigerian energy system. This considers the policy directions in terms of future energy demand and supply projections carried out by the Energy Commission of Nigeria (ECN). The assumptions made in this scenario are further discussed according to the Nigerian government plans for the household, commercial/service, transport, industry, agriculture, and power sectors as described in the national policies (NEMP 2014; NEP 2013; NREEE 2014).

In the Ref. scenario, the number of households is 37 million from the base year and this is expected to grow at 3.56 %. The increase in urban activities is expected to attract those in the rural areas and this will increase the urbanization rate to 50 % by 2040 from the 44 % in 2010. According to the NREEE (2014), the Nigerian government plans to introduce energy labeling program for households' energy-consuming appliances with the intention to reduce energy intensities. Also the encouragement of the widespread use of energy saving electric lamps such as CFL[24] and LED[25] will phase-out inefficient incandescent bulbs. This policy was not, however, considered in the development of this scenario due to the absence of a specific target, but it was developed for the alternative scenarios which will be discussed in the next subsections.

The National Bureau of Statistics reports that only 67 % of households in Nigeria depend on electricity supply from the national grid, the remaining 33 % depend on other sources which include private generators, rural electrification, etc. However, electricity access is expected to increase up to 75 % in 2020 and 100 % by 2030 as expressed in the NEMP (2014). To achieve this, the share of urban population without electricity is expected to shrink to 10 % by 2030 and 0 % by 2040, while rural communities will attain 80 % by 2030 and 100 % by 2040.

The NEMP (2014) stressed the need for the reduction of electricity use in cooking and heating, of which the government intends to supplement with cooking

---

[24] A compact fluorescent lamp (CFL), is designed to replace an incandescent lamp; some types fit into light fixtures formerly used for incandescent lamps. The lamps use a tube which is curved or folded to fit into the space of an incandescent bulb, and a compact electronic ballast in the base of the lamp. Compared to general-service incandescent lamps giving the same amount of visible light, CFLs use one-fifth to one-third the electric power, and last eight to fifteen times longer. A CFL has a higher purchase price than an incandescent lamp, but can save over five times its purchase price in electricity costs over the lamp's lifetime (Energy Star 2015).

[25] An LED lamp is a light-emitting diode (LED) product that is assembled into a lamp (or light bulb) for use in lighting fixtures. LED lamps have a lifespan and electrical efficiency that is several times better than incandescent lamps, and significantly better than most fluorescent lamps, with some chips able to emit more than 100 lumens per watt. The LED lamp market is projected to grow by more than 12-fold over the next decade, from $2 billion in the beginning of 2014 to $25 billion in 2023, a compound annual growth rate (CAGR) of 25 %. The efficiencies of the LED are expected to drive down the package costs (Jacques 2014).

gas stoves and solar thermal for water heating (Muhammad 2012; Shaaban and Petinrin 2014; Adekunle et al. 2015). The percentage share of fuel source for lighting (see Appendix) is assumed to remain the same until 2040, except for the case of rural electrification program in the rural areas (especially in non-electrified rural areas).

The transport, industrial and agriculture sector will continue to grow at its current growth rate, while energy consumption also follows the status quo. No improvement in energy efficiency is observed in the sectors, although the Nigerian government recognizes the need to improve the efficiency of other energy consuming sectors besides the household sector (Oyedepo 2012). Improvement in electricity supply to the industrial sector was not also considered in the Ref. scenario due to no clear policy direction in the current national energy policies. This implies that industries will continue to rely on private/independent power generating sets to support their activities.

On the power sector, the Nigerian government promoted some planning studies on the Nigerian electricity sector as a whole. This was done so as to present a sound understanding of the condition/challenges of the power sector and provide solutions. The studies include among others:

- The Presidential Advisory Committee Report on Nigeria's Electricity Sector in 2006 (see PAC 2006).
- The National Load Demand Forecast by the Power Holding Company of Nigeria Project Management Unit in 2009 (see PHCN-PMU 2009).
- The Nigerian National Energy Demand and Power Planning Study (2000–2030) by the Energy Commission of Nigeria (see ECN 2008).
- The Nigeria Vision 20:2020 Energy Report, Medium Term implementation Plan (2010–2013) by the National Planning Commission (see ECN 2008).

These reports presented some vital problems facing the power sector in Nigeria which include the poor state of transmission and distribution network in the national grid, inadequate electricity generation capacity, hence the low power output, inadequate supply of gas for power plants, etc.

This prompted the Nigerian government to set up the Presidential Action Committee on Power[26] (PACP) which is chaired by the President himself, and the Presidential task Force on Power (PTFP) which is chaired by the Advisor to the President on Power. The PACP presented a report which was titled "The Roadmap for Power Sector Reform (see FGN 2010) and this report along with the Nigeria Vision 2020 (ECN 2008) was the basis for the development of the future electricity supply projections in the Ref scenario.

In the development of the future electricity supply mix in the Ref scenario, the capacity addition will continue with the existing mix of thermal power and hydroelectric power plants for on-grid power, while diesel and gasoline power generators are provided for off-grid electricity generation. The fossil fuel electricity

---

[26]www.nigeriapowerreform.org/index.php?option=com_content&view=article&id=337&-Itemid=318.

generation technologies include a large share of natural gas of which single cycle power plants will have a 51 % share by 2040 and CCGT will have 24 %. The development is in line with Nigerian government's plan to increase electricity generation from gas power plants through the improvement of natural gas supply (Tallapragada 2009; Olugbenga et al. 2013). The Nigerian government intends to increase the prices of domestic gas supply, so as to be comparable with international gas prices (ICLG 2015).

Other electricity generating technologies such as nuclear is expected to have a 4 % share by 2040. This is expected to begin with the construction of a 1,000 MW power station by 2020 in Nigeria as proposed by the National Atomic Energy Agency (Lowbeer-Lewis 2010). Currently, the Nigerian government has selected two sites for the construction of a 4,800 MW nuclear power plant. The locations are Geregu in Kogi State and Itu in Akwa Ibom State, and each will have 2,400 MW at an investment cost of $20 billion.[27]

The minimal expansion for hydropower is expected to reach a capacity of 9,000 MW, while coal steam power plants will have a capacity of 13,000 MW by 2040. The Nigerian government has made consultations with relevant stakeholders and made plans to revive the coal mining industry.[28] This will supply the needed coal power plants under construction in various part of the nation (Odesola et al. 2013). In 2014, the construction of 1,000 MW coal power plant began in Enugu State,[29] while a 500 MW power plant is being built in 2015.[30] Operation of the coal power plants is expected to commence by the end of 2015,[31] while other power plants will become operational in the years to come.

Renewables such as on-shore wind turbines, solar thermal and PV systems are expected to have a small share in the future electricity generation mix. Off-grid technologies such as diesel, gasoline, and small gas turbines will increase from 4300 MW in 2010–47000 MW in 2040 due to the availability of fuel for powering these plants in remote areas of the country. The rest of the projected electricity generation capacity in Nigeria under the Ref scenario is show in Fig. 4.3 and Table A.1.

Transmission and distribution loss which stands at 20 % in the base year will be reduced to 10 %, while natural gas losses[32] will be reduced to 20 % by 2040.

---

[27]www.power-technology.com/news/newsnigeria-selects-sites-for-4800mw-nuclear-power-plants-4604192.

[28]www.bdlive.co.za/africa/africanbusiness/2013/06/05/nigeria-seeks-to-revive-mining-to-diversify-from-oil.

[29]www.vanguardngr.com/2014/09/fg-establish-1000mw-coal-power-plant-enugu.

[30]www.thisdaylive.com/articles/firm-to-build-500mw-coal-power-plant-in-enugu/202562/.

[31]www.punchng.com/business/business-economy/coal-power-plant-to-begin-operation-in-2015/.

[32]According to a study carried out by Achebe et al. (2012) to analyze the oil pipeline failures in the oil and gas industry, the losses were attributed to construction failures, external human material and structural accident, and incidental acts of vandalism and sabotage.

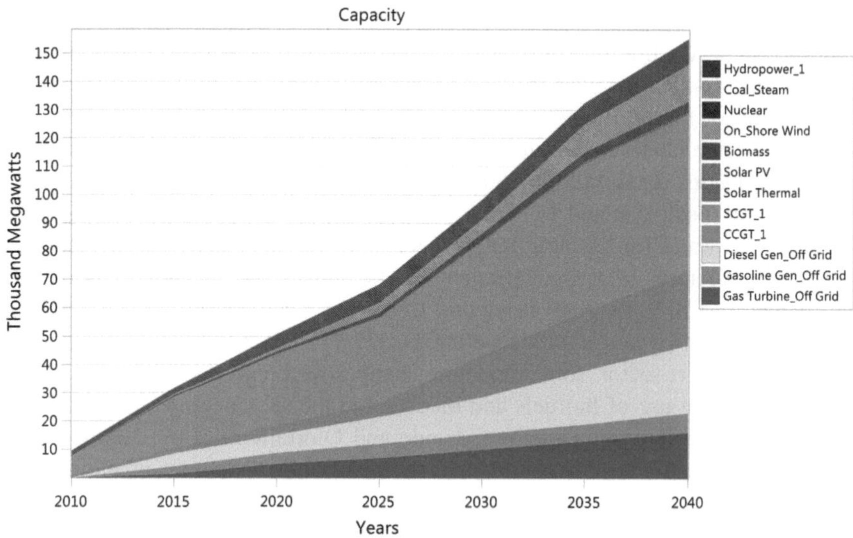

Fig. 4.3 Ref scenario future electricity generation capacity

Charcoal production in the Ref scenario is by traditional earth mound and brick kilns at 70 and 30 % respectively.[33]

4.3.3 The Low Carbon Moderate Scenario (LCM)

This scenario takes into consideration the need to improve power supply at a moderately reduced GHG emission as compared to the Ref scenario by 2040, and moderately reduced energy demand in the sectors. In setting up the low carbon scenarios (both LCM and LCA), the resource availability was observed for the continued supply of fuel for the power plants until 2040. However, imports of fuels such as uranium for nuclear power plants will commence in or before 2020 due to the construction of the nuclear power plants.[34] The low carbon technologies in the LCM are economically competitive with those on the Ref scenario. This includes energy technologies from the household sectors such as improved firewood and charcoal cooking stoves with lower intensities.

The lighting bulbs used in the household and commercial/services which are mainly incandescent lamps will be reduced to 15 % by 2040, while CFL will be

[33]This means that in the Nigerian LEAP model under the Ref scenario, charcoal is produced from 70 % traditional earth mound which is the common means of charcoal production in most part of Nigeria. The remaining 30 % comes from more modern brick kilns.

[34]This includes the import cost of nuclear fuel into Nigeria.

introduced in 2015 and reach 70 % by the end year. The CFL bulbs which cost about $2.33 have a life time of 6 years[35] which is more competitive than the incandescent bulbs which are sold for $0.33 and have a life time of 4–6 months.[36] The introduction of these efficient lighting bulbs will reduce the electricity consumption in commercial/service sector to 10 %, while households will have a drop in energy intensities for lighting by 2040. The CFL (14 watts) bulbs have the capacity to reduce peak demand (which is usually during the evening from 5 pm to 9 pm) by 46 watts for a single bulb, which is lower than the incandescent bulb (60 W). Refrigerators that are inefficient will be replaced by efficient ones and this will increase to 70 % by 2040 at the cost of $20 per refrigerator which will extend their lifetime to 14 years at a reduced intensity.[37]

In the transport sector, some moderate changes were introduced into the sector. This includes the use of biofuels and biodiesel in the six categories of vehicles in the Nigerian transportation sector (Akande and Olorunfemi 2009; Highina et al. 2011). However, there exist some challenges with the production of biofuels for commercial use as discussed in Idusuyi et al. (2012), and Balogun (2015).

These challenges[38] could be alleviated with the increased support of the Nigerian government through the National Biofuel Policy (NNPC 2007). Biofuels which are produced from contemporary biological process from either agriculture or anaerobic digestion,[39] present a future of reduced GHG emission in Nigeria as seen in countries such as the USA and Brazil. Biodiesel can be used as a fuel for vehicles in its pure form or as a diesel additive to reduce the levels of particulates, carbon monoxides, and hydrocarbons from diesel-powered vehicles.[40] The shares assumed for biofuel use in the Nigerian transport sector under the LCM scenario up to 2040 are presented in Table A.2.

The transmission and distribution losses in electricity sector assume a reduction to 6 % by 2040. This is in line with the current trend of low T&D loss experienced in other developing countries which are becoming more advanced. The reduction in T&D losses will be due to the improvement in the national grid network and provision of distributed generation and distribution systems (Anumaka 2012). Natural gas losses will also be reduced to 13 % in the LCM scenario, so as to ensure improved supply to all the sectors of the economy.

In order to ensure a reduction in the amount of GHG emission from electricity generation, the inclusion of renewable energy technologies was made in this scenario. Due to the high potentials for solar radiation in most part of Nigeria, the share of solar PV systems alone was assumed to be 9 % by 2040 (Yohanna and

[35]www.efi.org/factoids/cfl_faq.html.

[36]www.consumerenergycenter.org/lighting/bulbs.html.

[37]www.homeguides.sfgate.com/expected-life-refrigerator-88577.html.

[38]Some of the challenges include feedstock availability, land availability for feedstock production, availability of domestic market, and lack of support for the development of the biofuel industry.

[39]www.businessdictionary.com/definition/biofuel.html.

[40]www.userpages.umbc.edu/~copher1/ART336/wikiredesign/biofuel.html.

Umogbai 2010; Awogbemi and Komolafe 2011; Ogunmodimu 2013; Oghogho 2014). Rural electrification with solar PV systems and small hydropower was integrated for off-grid electricity supply as unlimited potentials exist in its application (Muhammad 2012; Shaaban and Petinrin 2014).

According to the study carried out by United States Environmental Protection Agency (USEPA 2010) and United Nations Industrial Development Organization (UNIDO 2011), a large amount of energy could be realized from biomass and small hydropower sites in Nigeria. The studies clearly showed that a 250 MW of biomass power plant constructed in Nigeria could generate 1,643 MWh/y of electricity in 2015 alone. This figure could go higher to 13,140 MWh/y of power supply with the installed capacity of 2000 MW in the long term. Also small hydropower plantwith an installed capacity of 100 MW can generate 526 MWh/y of electricity in 2015, and 17,870 MWh/y on the long term (see Table 2.7).

The increase in the installed capacity of CCGT in the LCM scenario is due to the lower carbon intensity, improved efficiency, availability time, and output capacity as compared to the SCGT. The share of off-grid gas plants was increased due to economical nature of gas power plants as compared to diesel and gasoline generators. Although gas supply may not get to some power plants located in remote areas of the country, diesel and gasoline may be available, but at a higher cost (Elusakin et al. 2014). This was also basis for the increased capacity of small hydropower and solar PV systems in off-grid areas.

The total capacity for the LCM scenario is 170,500 MW by 2040, which is more than the Ref scenario (i.e. 155,283 MW). Nuclear power plants are assumed to have an installed capacity of 5000 MW which is realizable and judging from the fact that a 4600 MW nuclear power plant is set to be constructed before 2020 (as discussed in the Ref scenario). Coal power plants were of two types in the LCM scenario; the conventional coal steam turbine and the more advance coal supercritical power plant. The coal power plants had an installed capacity of 25000 MW, of which the coal steam had 14000 MW and the coal supercritical had 11000 MW by 2040.

The basis for the selection of these two coal power plants was based on the economic variability assessment carried out by Ujam and Diyoke (2013) and the review by Essien and Igweoonu (2014). The projected generation capacity additions by technologies in the LCM scenario are presented in Table A.4 and Fig. 4.4.

The increase in crude oil refinery and coal mining production capacity was not assumed in this scenario, hence it remains the same as in the Ref scenario. However, charcoal production efficiency was improved through the introduction of steel kilns (40 % share) and casamance kilns (30 % share), so as to boost domestic production and meet market demand (see Table A.3).

4.3.4 The Low Carbon Advance Scenario (LCA)

In making a shift towards a more advanced low carbon society in Nigeria, the LCA scenario was developed. This scenario assumes a more aggressive energy

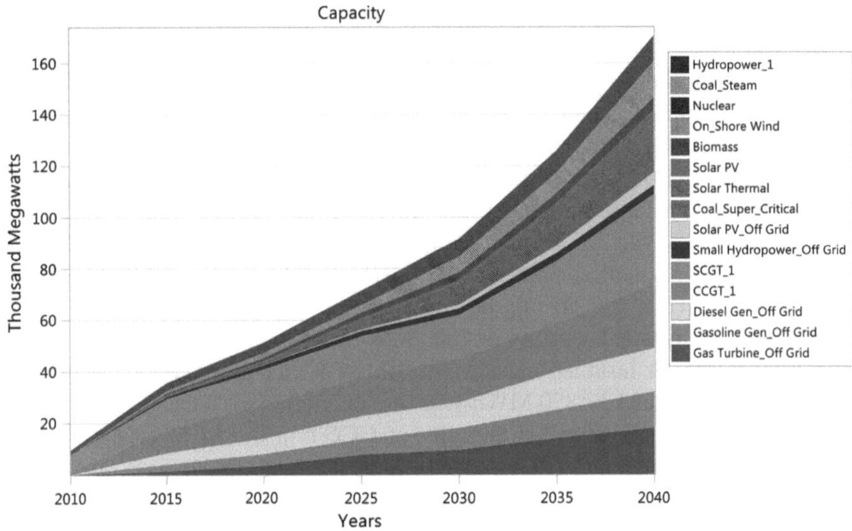

Fig. 4.4 LCM scenario future electricity generation capacity

policy and strategy to reflect a higher level of economic development and commitment for climate-friendly energy production and consumption than the LCM scenario. As shown in Table 4.1, the LCA scenario has more sectorial growth rate as compared to the Ref and LCM scenarios. The assumption for the annual economic growth is much higher than the average annual growth rate and much higher than the projections made by the National Bureau of Statistics and the Central Bank of Nigeria.

In the household sector, the increase in income is assumed to increase the purchase of more household electrical appliances such as electronics (TV, DVD, Hi-Fi systems), air conditioners, and refrigerators. With this, household energy consumption is expected to increase more than the Ref scenario. To address the increase in energy demand and reduce household energy intensities for electrical appliances, energy efficiency labeling program for the Nigerian household is implemented. This includes the increased ownership of efficient refrigerators, air conditions, and other household appliances. This policy's strategy is expected to bring about the reduction of household energy intensity to about half the required amount by 2040 (Momodu et al. 2012).

Efficient lighting in the household sector in the LCA scenario is expected to attain much lower energy intensity through the introduction of LED bulbs which are more efficient than the CFL bulbs introduced in the LCM scenario. The LED[41] bulbs consume about 30 % less energy than the CFL bulbs and about 50 % less than the incandescent bulbs. Comparing the LED, CFL, and incandescent light bulbs, the LED bulbs present a better alternative option in terms of lower annual

[41]www.bulbs.com/learning/ledfaq.aspx.

Table 4.3 Comparison between LED, CFL, and Incandescent light bulbs

Energy efficiency and energy costs			
	LED	Incandescent light bulbs	CFL
Life Span (average) (h)	50,000	1,200	8,000
Watts of Electricity used (equivalent to 60 W bulb) (W)	6 – 8	60	13 – 15
Kilo-watts of Electricity used (30 Incandescent bulbs per year) (KWh/yr)	329	3285	767
Annual Operating Cost (30 Incandescent bulbs per year equivalent)	$32.85/year	$328.59/year	$76.65/year
Environmental impact			
Toxic mercury	No	No	Yes
RoHS compliant	Yes	Yes	No
Carbon dioxide emissions (30 bulbs per year) (lb/year)	451	4500	1051
Light outputs			
Lumens	Watts	Watts	Watts
450	4 – 5	40	9 – 13
800	6 – 8	60	13 – 15
1,100	9 – 13	75	18 – 25
1,600	16 – 20	100	23 – 30
2,600	25 – 28	150	30 – 55
Other facts			
Sensitivity to low temperatures	None	Some	Yes
Sensitive to humidity	No	Some	Yes
On/off cycling	No effect	Some	Yes
Turns on instantly	Yes	Yes	No
Durability	Very Durable	Not very durable	Not very durable
Heat emitted	3.4 btu's/h	85 btu's/h	30 btu's/h
Failure modes	Not typical	Some	Yes

Source www.designrecycleinc.com/led%20comp%20chart.html

operating cost, intensity, environmental impact, output, and life span as shown in Table 4.3. Although LEDs cost $4.5 more than the CFL and incandescent bulbs in Nigeria, based on its benefits as shown in Table 4.3, it is expected to offset the cost in the long run.

The share of LPG for lighting and cooking increased to 20 %, while the share of kerosene reduced to 11.70 % in the LCA scenario. This is expected to be a better alternative than kerosene which has been in high demand by the household sector in Nigeria.[42] A study by Maduka (2011) showed that the popularization of LPG among Nigerian women can help in the reduction of firewood use in Nigerian households. The challenges however remain the access to the LPG in rural areas and lack of public awareness on the use of LPG in households (Ah Julius 2013). LPG has an important role to play in achieving a low carbon economy in Nigeria and other developing countries (Kojima 2011; Kojima et al. 2011), hence the increase in its share was considered in the LCA scenario.

According to Nonekuone (2008), the barriers to the increase use of LPG in Nigeria include affordability, pricing, government policies, safety, transportation, and distribution. These challenges were assumed to be addressed by the government through the government provision of policy regulation, institutional development, safety standards, and improved access to the rural areas. Besides the reduction in the share of firewood and kerosene use for cooking, more policy steps were taken in improving the efficiency of firewood and kerosene cooking stoves in the rural and urban areas in the LCA scenario. These steps are expected to lower the demand for these fuels in the household sector which is expected to have the highest consumption in the Ref scenario.

Efficiency improvement through the increased use of LEDs does not only affect the household sector, but also the commercial/service, and industrial sector. Further improvement in energy efficiency in the industry sector is possible through replacement of inefficient electric motor, air conditioning, and refrigeration systems. According to the energy audit carried out on manufacturing and processing industries in Nigeria by Oyedepo and Aremu (2013), electrical energy consumption by electric motors accounted for about 40–47 % of the total electricity consumed in most industries in Nigeria. Others such as boilers and heaters accounted for about 65 % of the total energy consumed. Inefficient energy utilization in industrial sector in Nigeria was attributed to poor in-housekeeping of industrial machineries/equipment such as air conditioner, refrigeration equipment, electric motors, etc. and non-utilization of renewable energy sources such as natural light in the day time which is the result of the habit of not switching off electric bulbs during the day time. This inefficient utilization of energy, if addressed, will result in the reduction of about 30 % of energy demand in the industrial sector in Nigeria.

The use of LPG which is also called autogas when used as a fuel in internal combustion engines in vehicles is being extensively used in many countries (Sethiya 2014). Countries such as Australia, Italy, Poland, South Korea, and Turkey account for half of the world's autogas vehicle consumption in 2013.[43] Autogas Incentive Policies (AIP)[44] have been credited with the success of the

[42]www.thisdaylive.com/articles/hindering-growth-of-nigeria-s-lpg-market-with-kerosene-subsidy/157397.

[43]See www.bpnews.com/index.php/publications/magazine/current-issue/305-world-autogas-demand-jumps-57-in-past-decade-incentives-play-key-role.

[44]The AIP is promoted by the government through lowering fuel tax vis-à-vis gasoline and diesel, lowering vehicle taxes or conversion subsidies, employing traffic measures, and removing barriers (parking restrictions, etc.).

increased use of LPG for transportation in most countries. According to the World LP Gas Association (2014), government policies to promote alternative fuels such as autogas have been mainly through financial incentives, regulatory policies and measures. These were assumed to be ineffective in the LCA scenario and therefore ensure the penetration of LPG (or autogas) vehicles in Nigeria by 2040. The shares of LPG cars for vehicles in the transport sector under the LCA scenario are shown in Table A.5. The assumption was made in line with penetration of LPG in other developing countries such as China as reviewed by Leung (2011), and Ou et al. (2010), with the scenario analysis for LPG penetration in Ghana by Biscoff et al. (2012).

The LCA scenario assumes more expansion of renewable energy technologies in both rural and urban areas in order to increase electricity supply, energy security, and enhance environmental protection as compared to the LCM scenario. To meet the environmental obligation of reduced GHG emission, the LCA scenario assumes a strong political pressure to build power plants that are less polluting than those in the Ref and LCM scenarios. Table A.6 and Fig. 4.5 present the projected electricity generation capacity for the LCA scenario. From the generation mix, diesel and gasoline generators in the off-grid areas are put to retirement and never expanded as in the Ref and LCM scenarios. They are, however, replaced with more efficient gas turbines and small SCGT power plants at 25 and 27 % respectively.

Renewables for off-grid electricity generation are also improved at 15, 17 and 16 % for on-shore wind, small hydropower, and solar PV respectably. The share of renewables also increased for on-grid electricity generation with solar PV and large hydropower having 10 % share each. However, the more efficient CCGT

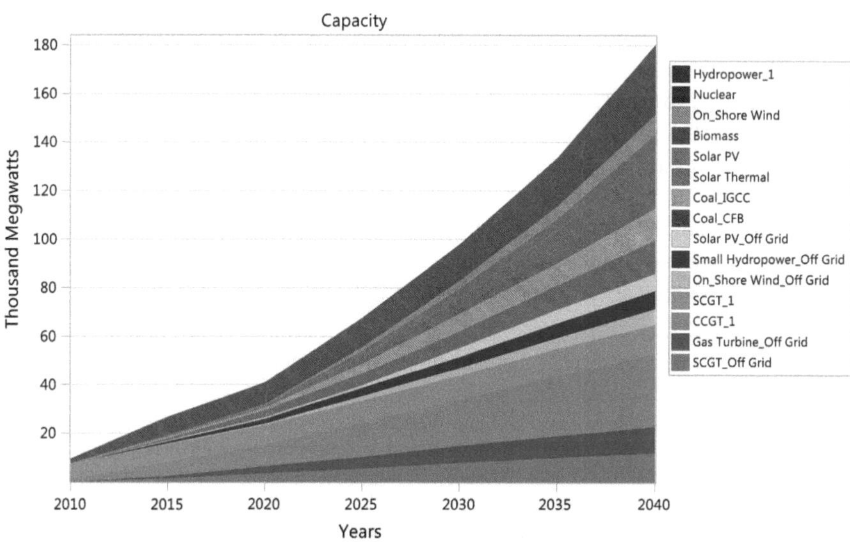

Fig. 4.5 LCA scenario future electricity generation capacity

power plants have 22 % of the total on-grid electricity generation capacity. Nuclear power plants share was also increased from 4 % in the LCM scenario to 11 % so as to ensure a more low carbon approach to electricity generation. According to the Nigeria Atomic Energy Commission (NAEC) and the Nigerian Nuclear Regulatory Authority (NNRA), the first nuclear power in Nigeria will have the capacity to produce 1,000 MW and be expanded to 4,000 MW within 10 years of establishment.[45]

Coal power plants in the LCA scenario are circulating fluidized bed (CFB) and integrated gasification combined cycle (IGCC) which are more efficient and produce less GHGs as compared to other coal power plants. Transmission and distribution losses in electricity reduced to 5 %, while natural gas pipeline losses reduced to 10 %. For charcoal production, Mud beehive and Adam retort kilns are used with a share of 40 and 35 % respectively. Oil refinery capacity was kept at 90 % for all the three refineries (i.e. Kaduna, Warri, and Port Harcourt Refineries).

4.3.5 The Green Optimistic Scenario (GO)

According to the National Renewable Energy and Energy Efficiency Policy (NREEEP 2014), Nigeria's population is expected to be doubled in twenty years and the aggregate energy demand will triple. Fossil fuel energy resources alone will not be able to meet the challenges of an increasing population at affordable costs and in a flexible manner (NEP 2013). In order to meet the rapidly growing demand for energy and the challenges posed by climate change, there has to be a conscious effort to increase the share of renewables in the energy mix (Oyedepo 2012; NREEEP 2014).

Great potentials and benefits exist for renewable energy development in Nigeria of which it can create jobs such as green jobs[46] (Bowen 2012), alleviate poverty as well as energy poverty (Emodi and Boo 2015a), and open up market in rural areas (REP 2015). These were the bases for the development of the GO scenario which intends to increase the share of renewables on the supply side, as well on the demand side.

On energy efficiency measures in the household and commercial/service sector, this scenario assumes that incandescent bulbs will be completely replaced by CFC and LED at 60 and 40 % respectively by 2040. To ensure that the GO scenario

[45]www.punchng.com/news/nigerias-first-nuclear-power-plant-ready-in-10-years-naec.

[46]The green jobs, green growth or the green revolution has been encouraged by many economies of the world such as South Korea. The South Korean government have been on the forefront of green growth initiatives and established the National Strategy for Green Growth (NSGG) (2009–2050) and the Five-Year Plan (FYP) (2009–2013) to provide a comprehensive policy framework for green growth in both the short and long term. The long-term policy impact of the NSGG aims to promote eco-friendly new growth engines, enhance peoples' quality of life, and contribute to international efforts to fight climate change. The FYP outlines the government's plan to spend approximately 2 % of annual GDP on green growth programs and projects (see OECD 2011).

achieves this, measures such as the "Phase out Incandescent Light Bulbs (POILB)" should be employed.[47] This measure comes as a kind of regulation effectively banning the manufacture, importation, and sale of incandescent light bulbs to the public.[48]

Most countries around the world have employed this policy measures including China that started the ban in 2012, but extended it to 2016[49]; India also started the ban on incandescent bulbs in 2012[50]; Israel phased it out since 2012[51]; the United Kingdom in 2011[52]; and all European Union (EU) countries to completely phase out incandescent bulbs before 2016.[53] Other countries include Canada who made a move in 2007,[54] while most states in the United States (US) completely phased out incandescent since 2007 and the others will in 2018.[55] Some countries implemented new energy standards and phased out incandescent bulbs, among them are Argentina in 2012,[56] Mexico, Malaysia and South Korea[57] in 2014.

Other efficiency measures include the complete deployment of efficient refrigerators and air conditioners which will phase out the inefficient ones used in the Nigerian households by 2040. Since the GO scenario is renewables oriented, the share of kerosene, firewood, and charcoal stoves is reduced to half the value in the LCA scenario. This is complemented by increasing the share of LPG and introducing solar thermal and solar cookers. Agbo and Oparaku (2006) reviewed the status and prospect of solar heaters in Nigeria. They highlighted some pressing policy issues, research and development (R&D), pilot and demonstration projects, institutional framework, investment promotion, incentives and protections. The GO scenario assumes that these measures will be met before 2040.

Alternative source of electric lighting for households and commercial/service sector encouraged with LPG use increased to 30 %, kerosene reduced to 10 %, solar lamp increased to 10 %, electricity stands at 48 %, other sources were 2 %. Industrial efficiency increased by 40 % with the same measures applied in the LCA scenario. The introduction of electric arc blast furnace will replace the

[47]See www.energyrating.gov.au/products-themes/lighting/lighting-and-phase-out-general-information/incandescent-light-bulbs-phase-out/.

[48]www.lampochki.org.ua/en/topovaja-novost-4/.

[49]www.reuters.com/article/2011/11/05/us-china-light-bulbs-idUSTRE7A40MV20111105.

[50]www.treehugger.com/interior-design/india-to-phase-out-400-million-incandescent-lightbulbs-by-2012-replace-with-cfls.html.

[51]www.energy.gov.il/LightBulb/Pages/GxmsMniMiniSiteLightBulb.htm.

[52]www.news.bbc.co.uk/2/hi/uk_news/7016020.stm.

[53]www.greenpeace.org.uk/blog/climate/eu-ban-inefficient-light-bulbs-eventually-sort-20081212.

[54]www.reuters.com/article/2007/04/25/us-lightbulbs-env-idUSN2529253520070425.

[55]www.leginfo.ca.gov/pub/07-08/bill/asm/ab_1101-1150/ab_1109_bill_20070223_introduced.html.

[56]www.lanacion.com.ar/1091978-desde-2011-no-podran-venderse-mas-lamparas-incandescentes.

[57]www.yonhapnews.co.kr/economy/2013/07/16/0302000000AKR20130716057151003.H-TML.

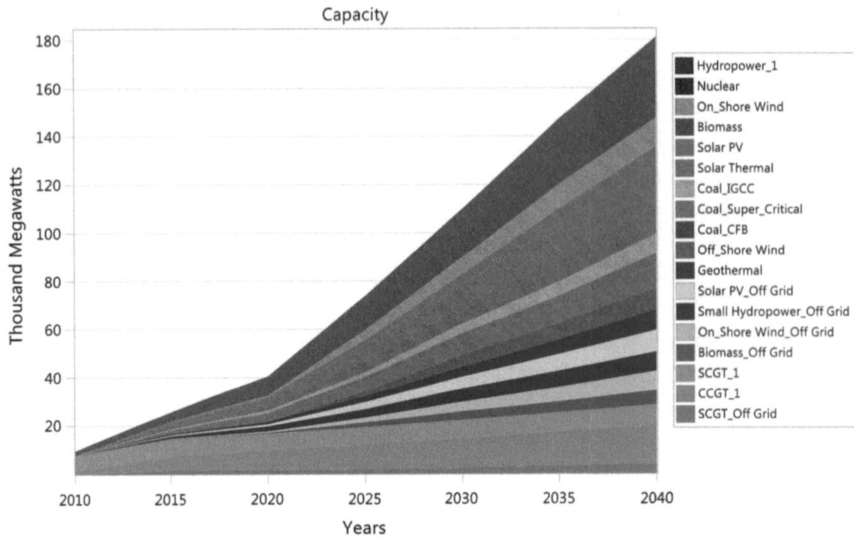

Fig. 4.6 GO scenario future electricity generation capacity

conventional blast furnace by 2040.[58] The use of the electric arc blast furnace in the Nigerian industries will not only reduce energy/electricity consumption, but also reduce GHG emission from the industrial sector. Due to the introduction of the electric arc, the share of energy consumed in the in the industrial sector is assumed to be as follows: electricity 43.94 %, biomass 23.46 %, natural gas 29.56 %, residual oil 2.735 and coal 0.32 % by 2040. For the growth rate of the industry, commercial/service and agriculture sector in the GO scenario, please refer to Table 4.1.

The transport sector inherits the biofuel and LPG used in the LCM and LCA scenarios, with the introduction of the CNG in all the vehicle categories apart from motorcycles. The GO scenario assumes the increased penetration of biodiesel use in motorcycles up to 50 % by 2040. In order to provide various low carbon emitting fuels than conventional ones, CNG and LPG are introduced and some vehicle types have the same share, while others slightly differ. Other policies introduced include improved fuel economy and reduction in private car ownership. The projected share of fuel mix in the transport sector under the GO scenario is presented in Table A.7.

In the power sector under the GO scenario, the share of fossil fuel power plants was reduced in terms of installed capacity as shown in Fig. 4.6 and Table A.8, except for CCGT power plants which have 11 % share in the total

[58]An electric arc furnace is a furnace that heats charged material by means of an electric arc. It is used for steelmaking and has various advantages over the conventional blast furnace which include the high efficiency in recycling of steel scrap and the efficient energy consumption of the arc.

on-grid electricity generation mix. Nuclear power plants have about 18000 MW of installed capacity while hydropower has 11 % share.

As observed from Fig. 4.6, the installed capacity of renewables in GO scenario is more than the LCA scenario, with the introduction of new renewable technologies: geothermal and off-shore wind turbine. The Nigerian energy policies (NEMP 2014; NEP 2013) stated the intention of the Nigerian government to investigate other source of renewable energy such as hydrogen and geothermal energy resources. This scenario did not consider hydrogen as an energy source, but assumed the introduction and expansion of geothermal power plant (from 2020–2040). Studies (Nwachukwu 1976; Avbovbo 1978; Gelnett and Gardner 1979; Onuoha and Ekine 1999; Nwankwo et al. 2009) have analyzed various parts of Nigeria for geothermal resources.[59] Kurowska and Krzysztof (2010) highlighted that the geothermal resources in Nigeria were enough to be exploited for power generation. This scenario takes this as a rationale for the inclusion of geothermal power plants in the future energy mix in Nigeria.

Off-shore wind was also introduced and expanded in the energy mix as a study carried out by Onyemechi et al. (2013) showed through a cost comparative assessment that off-shore wind power was possible and presents more advantage than other renewables in Nigeria coastal areas. For off-grid electricity generation, the status quo of retired plants in the LCA scenario holds, including the gas turbine power plants. The SCGT was the only fossil fuel plant considered for off-grid electricity generations at a small percentage share. Renewables such as the on-shore wind, small hydropower and solar PV are to complement the retired fossil fuel power plants. The GO scenario assumes a reduction of T&D losses to 4 %, while natural gas pipeline loss will be reduced to 6 % by 2040.

References

Achebe CH, Nneke UC, Anisiji OE (2012) Analysis of oil pipeline failures in the oil and gas sector. In: Proceedings of the international multiconference of engineers and computer scientists 2012, vol. II, IMECS 2012, Hong Kong. www.iaeng.org/publication/IMECS2012/IMECS2012_pp1274–1279.pdf

Adekunle OR, Jaafar AB, Musa MN (2015) The potentials of solar thermal technology for sustainable development in Nigeria

Agbo SN, Oparaku OU (2006) Positive and future prospects of solar water heating in Nigeria. The Pac J Sci Technol 7(2):191–198

Ah Julius A (2013) Households' access and preference to cooking fuels in Abuja, Nigeria. J Environ Sci Technol 6(2):91–98

Akande SO, Olorunfemi FB (2009) Research and development potentials in biofuel production in Nigeria. Afr Res Rev 3(3)

Akinyele DO, Nair NKC, Rayudu RK, Seah WKG (2014) Clean development mechanism projects for developing countries: Potential for carbon emissions mitigation and sustainable

[59]Maximum temperature gradients were found in around the Niger Delta regions within the sedimentary Basins, the Nupe Basin in west central Nigeria, Wikki Warm Spring area in the north-eastern part of Nigeria, the Anambra Basin.

development. In: 2014 Eighteenth National Power Systems Conference (NPSC), pp 1–6. IEEE

Anumaka MC (2012) Analysis of technical losses in electrical power system (Nigerian 330kV network as a case study). Department Of Electrical Electronic Engineering, Faculty of Engineering, Imo State University, Owerri, Imo State, Nigeria Email: engranumakamc@yahoo.com

Avbovbo AA (1978) Geothermal gradients in the southern Nigeria basin. Bull Can Pet Geol 26(2):268–274

Awogbemi O, Komolafe CA (2011) Potential for sustainable renewable energy development in Nigeria. Pac J Sci Technol 12(1):161–169

Balogun BO (2015) Potentials for sustainable commercial biofuels production in Nigeria. AFRREV STECH: Int J Sci Technol 4(2):25–40

Biscoff R, Akple M, Turkson R, Klomegah W (2012) Scenario of the emerging shift from gasoline to LPG fuelled cars in Ghana: a case study in Ho Municipality, Volta Region. Energy policy 44:354–361

Bowen A (2012) 'Green'growth,'green'jobs and labor markets. World bank policy research working paper (5990)

Cai W, Wang C, Chen J, Wang K, Zhang Y, Lu X (2008) Comparison of CO_2 emission scenarios and mitigation opportunities in China's five sectors in 2020. Energy Policy 36(3):1181–1194

Elusakin Julius E, Olufemi AO, Chuks DJ (2014) Challenges of sustaining off-grid power generation in Nigeria rural communities. Afr J Eng Res 2(2):51–57

Emodi NV, Boo KJ (2015) Sustainable energy development in Nigeria: current status and policy options. Renew Sustain Energy Rev 51:356–381

Energy Commission of Nigeria (ECN) (2008) Assessment of energy options and strategies for Nigeria: Energy demand, supply and environmental analysis for sustainable energy development (2000–2030), in Report No. ECN/EPA/2008/01, 2008. www.energy.gov.ng/index.php?option=com_docman&task=cat_view&gid=39&Itemid=49. Accessed 24 July 2015

Energy Star (2015) Light Bulbs for Consumers. http://www.energystar.gov/products/certified-products/detail/light-bulbs. Accessed 23 July 2015

Essien AU, Igweonu EI (2014) Coal based generation: a solution to Nigeria electricity problem. Int Archive Appl Sci Technol 5(1):74–80

Feng YY, Zhang LX (2012) Scenario analysis of urban energy saving and carbon abatement policies: a case study of Beijing city, China. Procedia Environ Sci 13:632–644

Gelnett RH, Gardner JV (1979) Use of radar for ground water exploration in Nigeria, West Africa. Mars, Incorporated

Heaps C (2006) Long-range energy alternatives planning system: data requirements for energy planning and mitigation assessment. Stockholm Environment Institute. www.energycommunity.org/documents/DataRequirements.pdf

Heaps CG (2012). Long-range energy alternatives planning (LEAP) system. [Software version 2014.0.1.18] Stockholm Environment Institute. Somerville, MA, USA

Highina BK, Bugaje IM, Umar B (2011) Liquid biofuels as alternative transport fuels in Nigeria. J Appl Technol Environ Sanitation 1(4):317–327

Howells M (2009) Contribution to the OSMOSYS forum open source energy modeling system. Working paper: documentation for OSMOSYS model 2009_1_1 (Draft). www.osmosys.yolasite.com/resources/OSMOSYS%202009_1_1.pdf

Howells M et al. (2011). OSeMOSYS: the open source energy modeling system: an introduction to its ethos, structure and development. Energy Policy 39(10):5850–5870

Ibitoye FI (2013) The millennium development goals and household energy requirements in Nigeria. SpringerPlus 2(1):1–9

Idusuyi N, Ajide OO, Abu R (2012) Biodiesel as an alternative energy resource in southwest Nigeria. Int J Sci Technol 2(5)

International Comparative Legal Guides (ICLG) (2015) Nigeria—Oil and Gas Regulation 2015. www.iclg.co.uk/practice-areas/oil-and-gas-regulation/oil-and-gas-regulation-2015/nigeria. Accessed 26 July 2015

Jacques C (2014) LED Lighting Market to Grow Over 12-Fold to $25 Billion in 2023, Lux Research. http://www.nigerianstat.gov.ng/ www.finance.yahoo.com/news/led-lighting-market-grow-over-123000345.html;_ylt=AwrBEiEyM_hSeA4AIC7QtDMD. Accessed 22 July 2015

Kojima M (2011) The role of liquefied petroleum gas in reducing energy poverty. www.wdronline.worldbank.com. Accessed 8 Aug 2015

Kojima M, Bacon R, Zhou X (2011) Who uses bottled gas? Evidence from Households in Developing Countries. Evidence from Households in Developing Countries (July 1, 2011). World Bank Policy Research Working Paper Series, Vol

Kurowska E, Krzysztof S (2010) Geothermal exploration in Nigeria. In: Proceedings world geothermal congress

Lazarus ML, Heaps C, Raskin P (1997) Long-range energy alternatives planning system—user guide for version 95. Stockholm Environment Institute, Boston

Leung V (2011) Slow diffusion of LPG vehicles in China—lessons from Shanghai, Guangzhou and Hong Kong. Energy Policy 39(6):3720–3731

Lowbeer-Lewis N (2010) Nigeria and nuclear energy: Plans and prospects. Nuclear Energy Futures Paper, (11). www.cigionline.org/sites/default/files/nuclear_energy_wp11-web.pdf

MADUKA EMJO (2011) Popularizing the use of Liquefied Petroleum Gas (LPG) as a substitute for fuel wood among women in Nigeria. In: Proceedings of the 3rd international conference of the african renewable energy alliance on renewable energy and gender

Momodu AS, Oyebisi TO, Obilade TO (2012) Modelling the Nigeria's electric power system to evaluate its long-term performance. In Proceedings of the 30th international conference of the system dynamics society

Muhammad U (2012) Rural solar electrification in Nigeria: renewable energy potentials and distribution for rural development. SOLAR2012_0332

National Bureau of Statistics (NBS) (2015) Characteristics of Nigerian Households. www.nigeriastat.gov.ng. Accessed 23 July 2015

National Energy Master Plan (NEMP) (Draft Revised Edition) (2014) Energy Commission of Nigeria (ECN). www.energy.gov.ng. Accessed 20 July 2015

National Energy Policy; Draft Revised Edition (NEP) (2013). Energy Commission of Nigeria (ECN). Abuja: Federal Republic of Nigeria. www.energy.gov.ng. Accessed 6 Aug 2015

National Renewable Energy and Energy Efficiency Policy (NREEEP) (2014) Energy Commission of Nigeria (ECN) and Federal Ministry of Science and Technology (FMST). www.energy.gov.ng. Accessed 6 Aug 2015

Nigerian National Petroleum Corporation (NNPC) (2007) Draft Nigerian bio-fuel policy and incentives. Nigerian National Petroleum Corporation, Abuja

Nonekuone J (2008) Model for marketing liquefied petroleum gas in Nigeria: Warri as a case study/Nonekuone Jolomi

Nwachukwu SO (1976) Approximate geothermal gradients in Niger Delta sedimentary basin. AAPG Bull 60(7):1073–1077

Nwankwo LI, Olasehinde PI, Akoshile CO (2009) An attempt to estimate the Curie-point isotherm depths in the Nupe Basin, West Central Nigeria. Global J Pure Appl Sci 15:427–433

Odesola IF, Eneje S, Olugasa T (2013) Coal development in Nigeria: prospects and challenges. Int J Eng Appl Sci 4(1):64–73

Oghogho I (2014) Solar energy potential and its development fortainable energy generation in Nigeria: a road map to achieving this feat. Int J Eng Manage Sci 5(2):61–67

Ogunmodimu OO (2013) CSP technology and its potential contribution to electricity supply in northern Nigeria. Int J Renew Energy Res 3(3):529–537

Ogwumike FO, Ozughalu UM, Abiona GA (2014) Household energy use and determinants: evidence from Nigeria. Int J Energy Econ Policy 4(2):248–262

Olugbenga TK, Jumah AGA, Phillips DA (2013) The current and future challenges of electricity market in Nigeria in the face of deregulation process. Afr J Eng Res 1(2):33–39

Onuoha KM, Ekine AS (1999) Subsurface temperature variations and heat flow in the Anambra Basin, Nigeria. J Afr Earth Sci 28(3):641–652

Onyemechi C, Igboanusi CC, Ezenwa EA (2013) A logistic analysis of Nigeria's offshore windfarm sector. Int J Res Soc Sci 4(4). www.ijsk.org/uploads/3/1/1/7/3117743/9_transport_management.pdf
Organisation for Economic Co-operation and Development (OECD) (2011) Towards Green Growth: A summary for policy-makers. www.oecd.org/greengrowth/48012345.pdf
Oteh OU, Agwu NM, Nwaogu EN, Nto CP (2015) Mitigating climate change and determinants of access to liquified petroluem gas (LPG) among urban households in Abia State, Nigeria. J Earth Sci Clim Change
Ou X, Zhang X, Chang S (2010) Scenario analysis on alternative fuel/vehicle for China's future road transport: life-cycle energy demand and GHG emissions. Energy Policy 38(8):3943–3956
Oyedepo SO (2012) Energy and sustainable development in Nigeria: the way forward. Energy Sustain Soc 2(1):1–17
Oyedepo SO, Aremu TO (2013) Energy audit of manufacturing and processing industries in Nigeria: a case study of food processing industry and distillation and bottling company. Am J Energy Res 1(3):36–44
Power Holding Company of Nigeria—Project Management Unit (PHCN-PMU) (2009) National Load Demand System—National Energy Development Project—Draft final report Volume 1—National Demand Load Forecast, Tractebel Engineering Suez and Omega Systems, p. 50 quoting National Bureau of Statistics, 2006, Core Welfare Indicator Questionnaire Survey
Presidential Advisory Committee (PAC) (2006) A draft report on Nigeria's electricity sector. Sub-Committee of the President Advisory Committee on 25 Years Electric Power Supply Projection, Abuja
Sethiya A (2014) LPG as an alternative fuel in automobile: a comparative analysis. J Appl Eng 2(1)
Shaaban M, Petinrin JO (2014) Renewable energy potentials in Nigeria: meeting rural energy needs. Renew Sustain Energy Rev 29:72–84
Stockholm Environment Institute (SEI) (2008) User guide, LEAP: long range energy alternative planning system. Stockholm Environment Institute, Boston
Tallapragada PV (2009) Nigeria's electricity sector-electricity and gas pricing barriers. International Association for Energy Economics (First Quater), 29–34
The Presidency—Federal Republic of Nigeria (FGN) (2010) Roadmap for Power Sector Reform (A Customer-Driven Sector-Wide Plan to Achieve Stable Power Supply). www.nigeriapowerreform.org/index.php?option=com_phocadownload&view=category&id=10:reform-documents&Itemid=352. Accessed 25 July 2015
The Renewable Energy Programme (REP), Federal Ministry of Environment of Nigeria (2015) Submissions to the United Nations Framework Convention on Climate Change (UNFCCC) on the LIMA Work Programme on Gender. www4.unfccc.int/submissions/Lists/OSPSubmissionUpload/39_122_130688171405397115-Nigeria%20gender.pdf
Ujam AJ, Diyoke C (2013) Economic viability of coal based power generation for Nigeria. Am J Eng Res 02(11):14–24
United Nations Industrial Development Organization (UNIDO) (2011) Independent UNIDO Country Evaluation: Federal Republic of Nigeria. www.unido.org/fileadmin/user_media/About_UNIDO/Evaluation/Country_Reports/Nigeria_E-Book.pdf
United States Environmental Protection Agency (USEPA) (2010) Landfill Recovery and Use in Nigeria (Pre-Feasibility Studies of Using LFGR). Final Report. Methane-To-Market Program, Washington DC 20460. www.globalmethane.org/Data/347_Landfill.Recovery.and.Use.in.Nigeria%28LFGE%29Final.Report.pdf
Webmeets (2015) Scenario cost-benefit analysis of German electricity market with new energy and environmental policies. www.webmeets.com/files/papers/eaere/2015/1089/Scenario%20costbenefit%20of%20Germany%20for%20EAERE%20conferencev1.pdf
World LP Gas Association (WLPGA) (2014) Autogas Incentive Policies: A country-by-country analysis of why and how governments encourage Autogas and what works (2014 Update). http://www.wlpga.org/wp-content/uploads/2015/09/autogas_incentive_policies_2014.pdf. Accessed 12 Aug 2015
Yohanna JK, Umogbai VI (2010) Solar energy potentials and utilization in Nigeria agriculture. J Environ Issues Agric Dev Ctries 2(2–3):10–21

Chapter 5
Results and Discussions

Abstract This chapter presents results of the Nigerian Energy Policies Strategies for Sustainable Development using LEAP model from 2010 to 2040. The chapter is divided into six sections; the first section presents results of the forecasted energy demand and supply, electricity supply, GHG emissions, and energy balance. The second section provides an in-depth analysis of the interrelations between energy, environment, and the society in Nigeria from 2010 to 2040 under the four scenarios proposed. The third section presents the least-cost electricity generation for the developed scenarios, which includes the GHG emissions. The fourth section analyzes the cost-benefit of the developed scenarios. The fifth section discusses sustainable strategies which are in line with the scenarios developed in this study. Finally, the sixth section summarizes the chapter with policy implications.

Keywords Forecasted energy demand and supply · GHG emission reduction potentials · Scenarios' analysis · Least-cost electricity generation · Cost-benefit analysis · Sustainable strategies

5.1 Results

In order to ensure a well-defined interpretation of the results, a scenario-based approach is employed for result interpretation. In this section, all scenario projections are analyzed based on the sectors specified in Chap. 4. To further simplify the interpretation of the results, the results are presented in the following order: future energy demand and supply, electricity supply projections, GHG emission projections and energy balance analysis.

5.1.1 The Reference Scenario (Ref)

This scenario was developed considering the most-likely-developmental-pathways of the Nigerian Government. The results are presented as follows.

N.V. Emodi, *Energy Policies for Sustainable Development Strategies*,
Frontiers in African Business Research, DOI 10.1007/978-981-10-0974-7_5

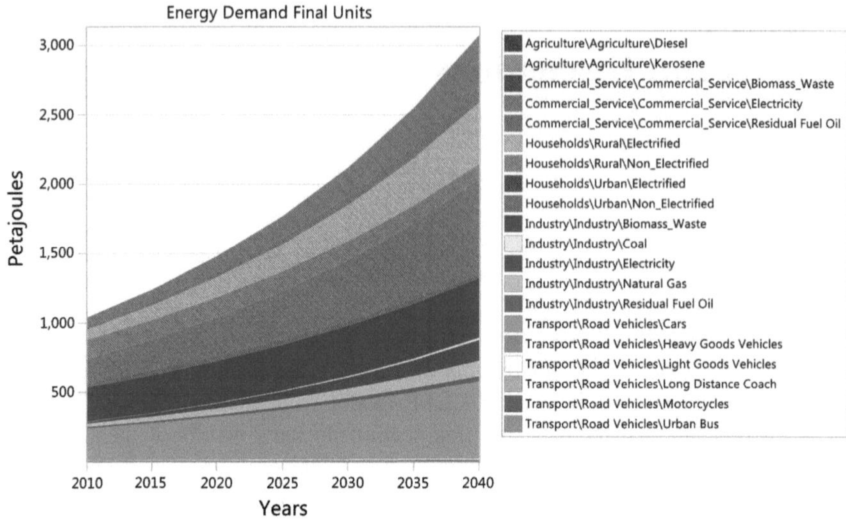

Fig. 5.1 Ref scenario energy demand

5.1.1.1 Future Energy Demand and Supply (Ref)

Future Energy Demand by Sector (Ref)

Energy demand under the Ref scenario is expected to grow from 1,039.3 PJ in 2010, to 3,075 PJ in 2040 as shown in Fig. 5.1 and presented in Table B.1. In the household sector, the urban household energy demand grows to 714.7 PJ in which the larger share of energy demand goes to the electrified households (i.e. 686.9 PJ). The rural households which are composed of both the electrified and non-electrified have a cumulative demand of 552.8 PJ by 2040. As can be observed from the results, the urban and rural electrified households' energy demand increased as compared to the non-electrified households in the urban and rural households.

This is due to the increase in population and income as can be recalled from Table 4.1 in Chap 4. Another reason for this feat in the household is based on the NEMP (2014) recommendations on the government's plan to achieve a 100 % electrification rate by 2040 and the need to reduce reliance on biomass for cooking and replace it with electricity in the household sector. This entails the need for the reduction in non-electrified households and the increase in electrified households.[1]

[1]This does not mean the physical reduction in household numbers, but the effect of switching the type of fuel used to meet energy needs (e.g. switching from firewood to electric stoves, while improving electricity access).

The industrial sector as observed in the results in the Ref scenario, presents a growing energy demand from 297 PJ in 2010 to 742.2 PJ by 2040. The larger share of demand by fuel remains biomass and waste which stands at 426.8 PJ in 2040 as against the 2010 value of 247.9 PJ. The share of electricity in the fuel mix is also expected to increase to 148.4 PJ by 2040, as is the case of natural gas which is 111.3 PJ. Other fuel sources such as residual fuel oil which is mainly used for powering privately owned electricity generator,[2] are assumed to grow at much lower rate as compared to other fuels. The growth in the energy demand in the industry sector is in line with the industrial growth rate specified in Table 4.1 in Chap. 4, and also in the absence of improved energy efficiency and electricity supply. In other words, the industries will continue to depend on privately owned generators to meet their electrical needs until 2040.

The commercial and service sectors which include public service sector and private business and service sector,[3] will have an energy demand of 482.5 PJ by 2040 and will compose of a larger share of biomass and waste (368.4 PJ) and electricity (112.7 PJ). The agriculture which according to the base year value of 0.1 and 0.3 for kerosene and diesel respectively, will have little increase in energy demand due to the low energy consumption experienced in the Nigerian agricultural sector.[4] In Nigeria today, the rate of mechanization has not improved for more than 30 years and this is due to the level of government's commitment in the development of the agriculture sector.[5]

The current status of vehicles in the transport sector under the Ref scenario assumed to value in the base year (see Tables B.15 and B.16), is expected to increase its total energy demand to 581 PJ by 2040. The fuel type which consists of gasoline and diesel is observed to be higher for cars with an energy demand increased from 229.6 PJ in 2010 to of 557.2 PJ in 2040. The reason for the increase in cars' demand is due to the increase in population size, as more people tend to buy more cars for private use and also the demand for fuel in cars with low/poor fuel efficiency. Motorcycles which are popular among the Nigerian populace for both private and commercial use, are expected to grow from 2.4 PJ in 2010 to 5.8 PJ[6] in 2040. The energy demand for other vehicles such as LGV, HGV, UB, and LDC is 4.9, 0.8, 4.1, and 8.1, PJ respectively.

[2]The electricity generated from privately owned generators is different from the electricity stated in the fuel mix, because the electricity in the fuel mix comes directly from the national grid supply.

[3]E.g. hospitals, small business enterprises.

[4]www.businessdayonline.com/2014/09/the-mechanisation-of-agriculture-in-nigeria/.

[5]This was the case presented in Lamidi and Akande (2013) study as the point that this was caused by the availability of capital, poor infrastructural facilities, poor attitude toward adoption of new innovation, and non-availability of storage means.

[6]Energy demand is measured in Petajoule (PJ) in order to ensure simplicity and comparison with other vehicles in the transportation sector. Its value has been converted from its previous unit which was KTOE.

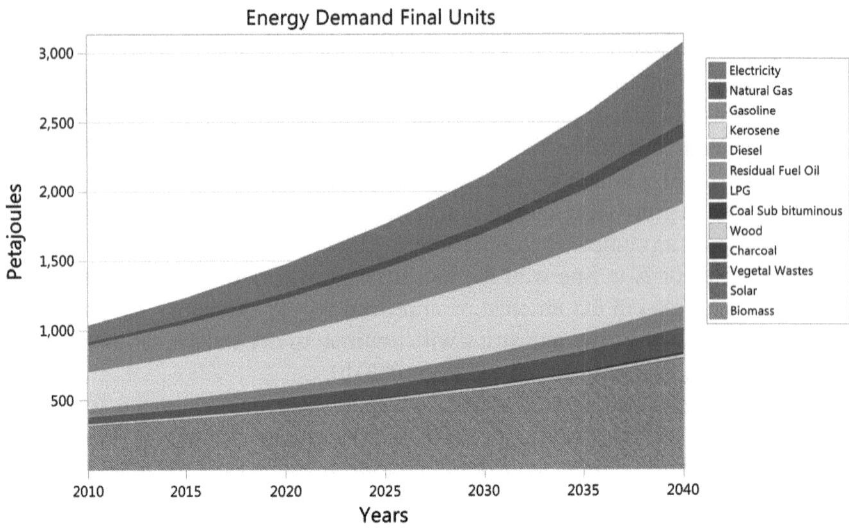

Fig. 5.2 Ref scenario energy supply

Final Energy Supply (Ref)

The final energy supply under the Ref scenario for both primary and secondary energy is shown in Fig. 5.2 and presented in Table D.2. The results show that out of the 3,075 PJ of the total energy supply, biomass (795.1 PJ) and kerosene (740.6 PJ) will be the most sought after fuel sources by 2040. Following this energy source will be electricity,[7] gasoline, LPG, diesel, and natural gas with the demand of 582.5 PJ, 470.4 PJ, 177.2 PJ, 112 PJ, and 111.3 PJ respectively. The increased demand of kerosene and biomass will be mostly concentrated in the household and industrial sector. As mentioned in the last subsection, the increase in the demand for energy is due to the considerations specified in Tables 4.1 and 4.2 in Chap. 4.

5.1.1.2 Future Electricity Supply (Ref)

From Fig. 5.3, the future electricity supply from the installed capacity of 155,283 MW in Ref scenario (Table A.1) is expected to generate 179,800 MWh by 2040. The power plant with the highest share of electricity production is the SCGT plants which will generate about 64,200 MWh, followed by CCGT power plants at 32,200 MWh, and diesel generators for off-grid power supply (30,400 MWh). In the Ref scenario, the government's plan to increase electricity access in off-grid areas will be achieve to a large extent through the use of diesel and gasoline

[7]The value of electricity here is measured in Petajoules (PJ) units which was converted from the Megawatts unit utilized/consumed in the sectors.

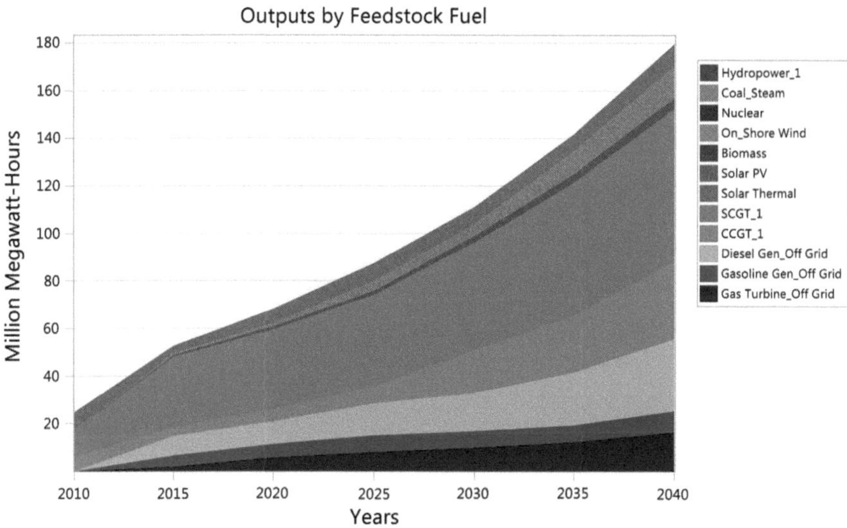

Fig. 5.3 Ref scenario electricity supply

generators, in association with small gas turbine (PTFP 2013). The nuclear power plant slated to come online in 2015 is expected to generate about 800 MWh to the population, and as expansion in its capacity grows, the power generation will contribute about 4,600 MWh by 2040.

The contribution of renewables in the Ref scenario is relatively small as compared to the supply from fossil fuel plants. The highest electricity supply of renewables will be generated from large hydropower plants[8] to the tune of 9,300 MWh, while other renewables such as on-shore wind, biomass, solar PV, and thermal will contribute about 16, 58, 170, and 22 MWh respectively by 2040.

5.1.1.3 GHG Emission (Ref)

The ever-increasing energy demand in the Ref scenario assures the rise in GHG levels from 42.4 million metric tons of carbon dioxide equivalent (MMTCDE)

[8]Although hydropower plants are actually renewables, most organizations and states in the United States of America and other countries don't consider larger hydropower plants as renewables and this due to the effects the dams has on aquatic habitat/life such as fisheries and water flows. This sometimes block migrating fish from reaching their spawning grounds. Further, dam reservoirs impact the flows, temperature and silt loads of rivers and streams (World Watch Institute [WWI] 2013; Daigneau 2013). Also in some countries such as Canada and those in the European Union, methane emissions have been credited to hydropower plants which is caused by the plant material in flooded areas decaying in an anaerobic environment, and forming methane, which is a GHG (World commission on Dams [WCD] 2000; Newscientist 2001). Other problems associated with hydropower plants includes the relocation of people living where the reservoirs are planned and the high rate failure risk due to poor construction, natural disasters and sabotage (International Rivers 2008; Osnos 2011; British Broadcasting Corporation News [BBC] 2012).

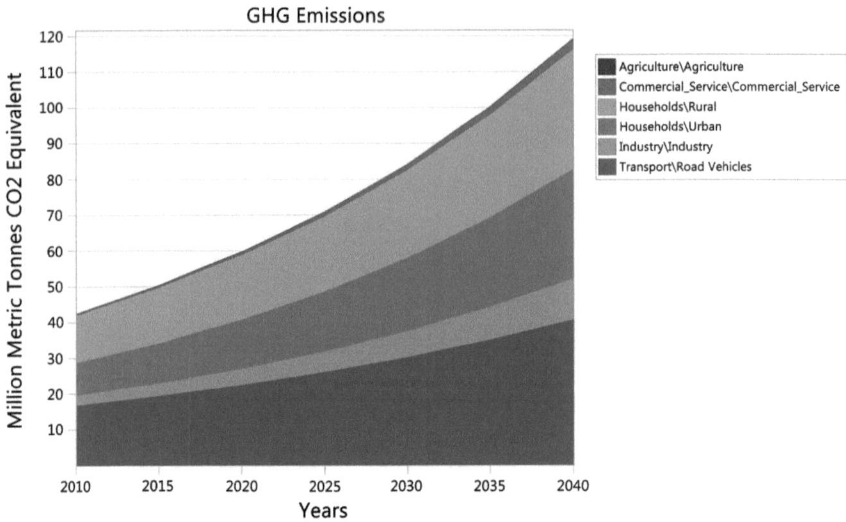

Fig. 5.4 Ref scenario GHG emission from energy consumption

in 2010 to 119.2 MMTCDE by 2040. The GHG emission under the Ref scenario is shown in Fig. 5.4 and presented in Table G.1. The increase as observed will have the highest contributions from three sectors: the households, industry, and transport sectors accounting for 64.1, 11.5, and 40.7 MMTCDE respectively. As described in the scenario development (please refer to Chap. 4), no meaningful policy strategy was in place to effect the reduction of GHG in all the sectors under this scenario. The GHGs from agriculture and commercial/service sectors will remain considerably small as compared to other sectors of the economy by 2040.

The GHGs from electricity generation were analyzed separately from the energy demand in the five sectors. In the Ref scenario, the fossil fuel power plants as shown in Fig. 5.5 and presented in Table G.5 are observed to emit 81.9 MMTCDE. Due to the capacity size of SCGT power plants in the electricity mix by 2040, the GHGs amount to about 32.3 MMTCDE and it is the highest among other fossil fuel power plants. The next power plant with high emission is the coal steam with an emission contribution of 15 MMTCDE by 2040. Off-grid energy technologies such as gasoline and small gas turbine will have less GHG emission of 6.6 and 5.5 MMTCDE by 2040.

5.1.1.4 Energy Balance (Ref)

The future energy balance in Nigeria was analyzed in order to examine the interactions between energy and the society between the year 2010 and 2040. In order to account for the changes in the energy system, the base year (2010) was represented using a Sankey diagram and this is shown in Fig. 5.6 and presented in Table H.1.

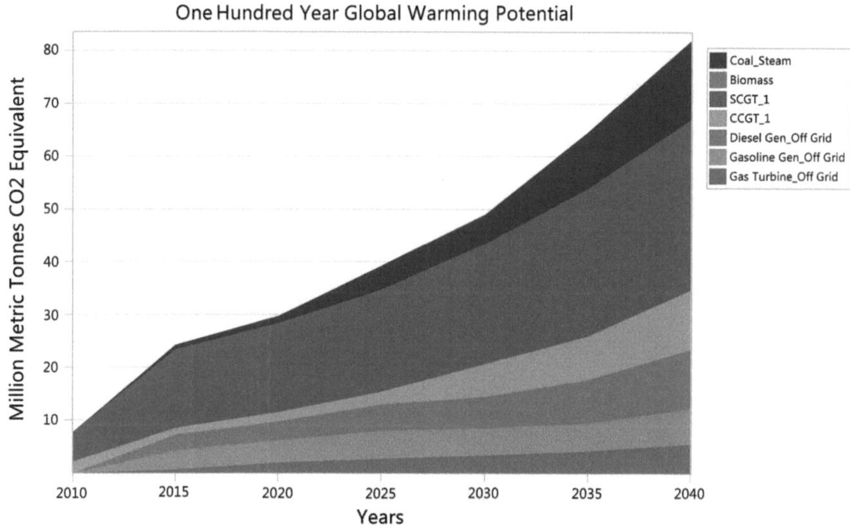

Fig. 5.5 Ref scenario GHG emission from electricity generation

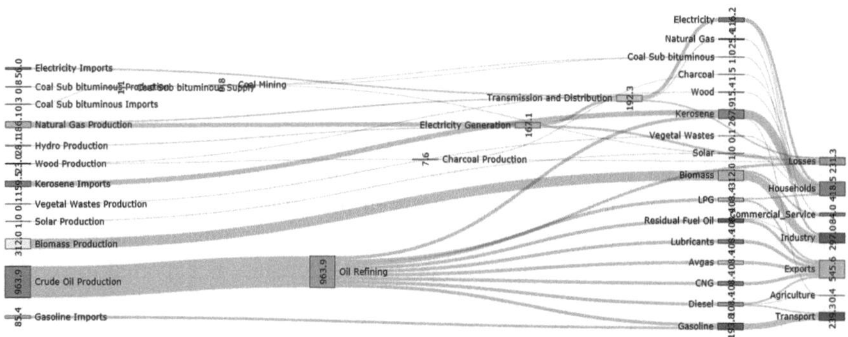

Fig. 5.6 Sankey diagram of Nigeria's energy balance in 2010 (Ref Scenario)

In the base year, the Ref scenario presents a situation whereby Nigeria will rely heavily on gasoline (85.4 PJ) and kerosene (159.5 PJ). According to the level of installed crude oil production capacity and products in the base year (Tables B.7 and B.18), the demand for natural gas which was 186.1 PJ was met from the local refineries in Nigeria. Out of this production (i.e. natural gas), 139 PJ was used for electricity generations in the gas power plants in 2010, while about 21.7 PJ was lost due to the process of transmission and distribution (T&D). Industrial energy demand for natural gas took the remaining share of 25.4 PJ.

Electricity generation in energy unit was 89.2 PJ and out of this was the loss of 29 PJ to T&D. Electricity supplied to the five sectors in which household electricity demand was 84.9 PJ, a deficit can be observed in the power supply as the

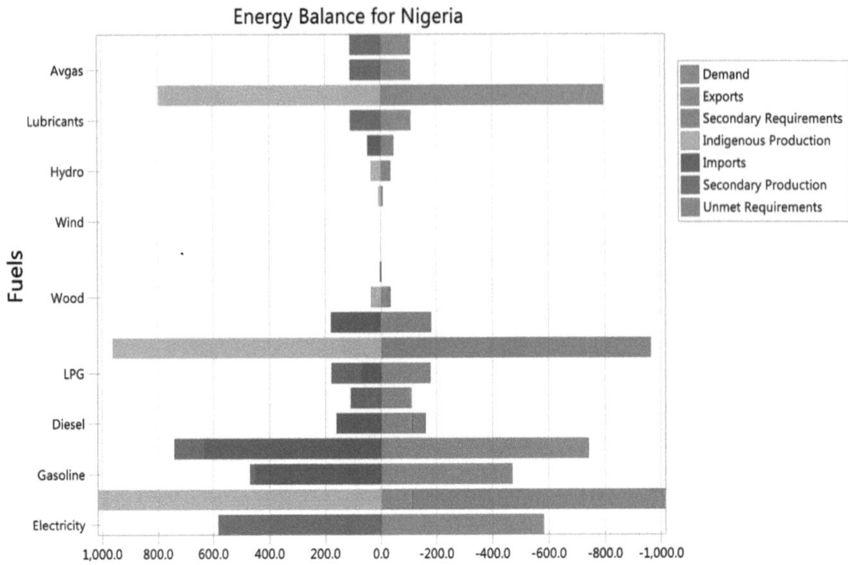

Fig. 5.7 Nigeria's energy balance in 2010 (Ref Scenario). (Order of fuels: CNG, Avgas, biomass, lubricant, hydro, solar, vegetal waste, charcoal, wood, coal sub-bituminous, crude oil, LPG, residual fuel oil, diesel, kerosene, gasoline, natural gas, electricity.)

production was unable to meet demand. In order to meet the electricity demand, the model assumed the importation of electricity[9] from near-by countries in the value of 56 PJ. The Sankey diagram also shows the interactions between the production and demand of other energy fuels in the Nigerian energy system. It is important to note that in this scenario, the production of a type of fuel that is not utilized in the domestic market or has no or less demand, is exported to foreign markets. This can be seen in some fuels such as diesel (62.6 PJ) which have less demand, residual fuel oil (97.1 PJ), LPG (60.6 PJ), lubricants (108.4 PJ), aviation gas (Avgas) and CNG (108.4 PJ).[10]

The chart showing the energy balance with provision for secondary energy requirements, indigenous production, secondary production, and unmet requirement is presented in Fig. 5.7. This chart was developed in order to further explore the Nigerian energy situation in 2010, as it could not be fully described in Fig. 5.6 and Table H.1. From Fig. 5.7, it can be observed that in 2010, the demand for biomass was met through indigenous production and this amounted to 312 PJ.

[9]In the real world situation, Nigeria does not import electricity from any neighboring country. However, in the Nigerian Model, the import can be seen as off-grid electricity production as clearly stipulated in Table A.1 and Fig. 4.3 of Chap. 4.

[10]It is also important to note that these fuels (except CNG) are actually in demand and used in Nigeria today, but due to the energy demand data for the various sectors (e.g. air transport for Avgas), this fuels were assumed to be exported in the base year.

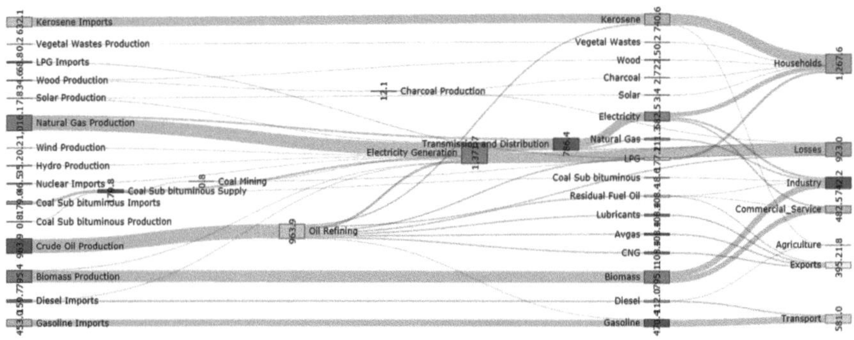

Fig. 5.8 Sankey diagram of Nigeria's energy balance in 2040 (Ref Scenario)

Imports had a considerable impact in meeting demand for kerosene and to an extent, gasoline in the base year under the reference scenario.

By 2040 under the Ref scenario, the Nigerian energy balance is expected to change, considering the fact that all sectors of the economy will experience growth increase, while the power sector will also have some improvement in the electricity mix. This is shown in Fig. 5.8 and presented in Table H.2. By 2040, Nigeria is expected to generate an energy equivalent of 582.5 PJ, which will be able to meet the demand in the households (321.4 PJ), industry (148.4 PJ) and commercial/service sector (112.7 PJ). The increase in fossil fuel power plants that utilize natural gas (e.g. SCGT, CCGT, gas turbine), will use 876.9 PJ of natural gas produced by 2040 for power generation, while 27.8 PJ will be lost due to T&D.

Gasoline, kerosene, diesel, LPG will be imported in a higher value than in 2010, to account for the increase in the households, transport and industry. In general, the total primary energy supply under the Ref scenario is 3,998 PJ in which 923 PJ is lost due to T&D, leaving about 3,075 PJ supply to the various sectors in Nigeria. Construction of the coal steam power plants with the installed capacity of 13,000 MW by 2040 is expected to use about 161.1 PJ of the 179 PJ imported sub-bituminous coal.[11] Renewables such as wind (0.2 PJ) and solar (7.8 PJ) will be used to power the solar and wind power plants in Nigeria.

Figure 5.9 graphically presents the Nigerian energy balance for 2040. The Figure shows that the demand for nuclear energy supply will be met through importation, and this will be for the 4000 MW capacity nuclear power plants which are expected to utilize about 46.5 PJ of nuclear fuel.[12] Gasoline energy

[11]In this case, the model assumes that the current state of coal mining in Nigeria will not improve due to the Nigerian government policies that have not considered the development of the coal mining sector that has been long dead on its feet since the 1960s.

[12]In this case, the nuclear fuel imported will be uranium. It is important to note that Nigeria has a considerable amount of uranium deposits in Cross River State, Adamawa State, Taraba State, Plateau State, Bauchi State, and Kano State as recorded by the British Geological Survey (WISE Uranium Project 2015).

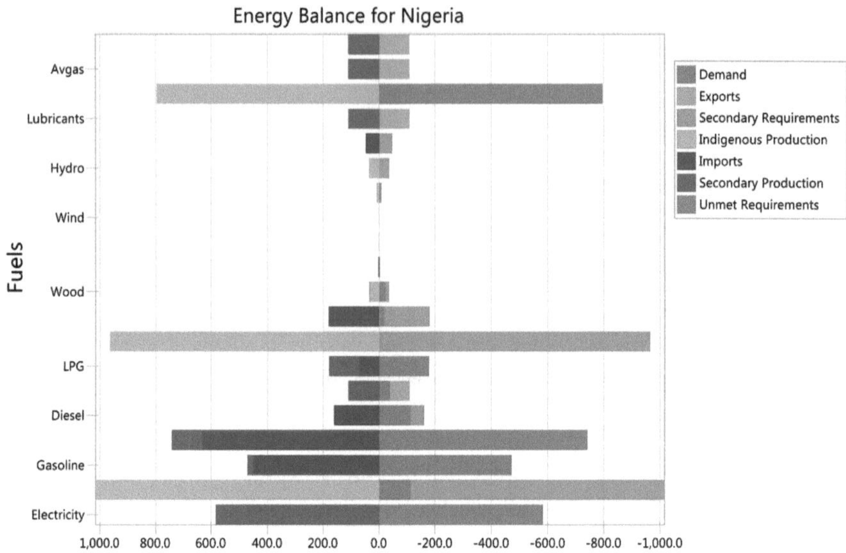

Fig. 5.9 Nigeria's energy balance in 2040 (Ref Scenario). (Order of fuels CNG, Avgas, biomass, lubricant, nuclear, hydro, solar, wind, vegetal waste, charcoal, wood, coal sub-bituminous, crude oil, LPG, residual fuel oil, diesel, kerosene, gasoline, natural gas, electricity.)

demand will be supplied mainly from imports (453 PJ), while secondary production will account for 108.4 and off-grid gasoline generators will use 91.1 PJ of it.

LPG use is expected to increase in the household sector as more people use LPG for cooking and this will require the importation of additional 68.8 PJ of LPG to complement the 108.4 PJ produced domestically. An alarming increase in kerosene use will also require the importation of 632.1 PJ to the domestically produced product, so as to meet the demand in the household (740.2 PJ) and agriculture (0.4 PJ) sectors by 2040. Diesel will mainly be imported by 2040 to meet both the demand such as transport (110.5 PJ) and agriculture (1.4 PJ), and secondary requirement such as for power generation (156.1 PJ).

5.1.2 The Low Carbon Moderate Scenario (LCM)

This scenario is driven by cheaper capital, cost and readily available fuels to improve power supply, and moderately reduces energy demand is the sectors analyzed. The results are presented as follows;

5.1.2.1 Future Energy Demand and Supply (LCM)

Future Energy Demand by Sector (LCM)

The quest to moderately reduce energy demand, while considering cheaper capital investment on various energy technology and policies, leads to the development of the LCM scenario. The results are shown in Fig. 5.10 and presented in Table C.2. The results show that the total energy demand in this scenario was 2,940.5 PJ as against the Ref scenario's energy demand of 3,075 PJ. The household sector's energy demand by 2040 was 1,157.4 PJ, which shows reduction from the Ref scenario. The non-electrified urban households were completely connected to the national grid by 2040, while little changes in the rural households were observed. However, the changes in the households energy demand are also in line with the scenario specifications in Tables 4.1 and 4.2 in Chap. 4.

The industry also experienced a reduction in energy demand from 742.2 PJ in the Ref scenario to 638.6 PJ in the LCM scenario. The only policy observed for the industry sector in this scenario was the general introduction of CFL bulbs for lighting which covers the household, commercial/services and industry sector. This is observed to have an impact in terms of reduction in electricity demand from 148.4 PJ in the Ref scenario to 127.7 PJ in the LCM scenario. In the commercial sector, CFL bulbs introduced were expected to reduce electricity demand but due to the impact of economic activity (e.g. GDP) may have contributed to the increase in electricity demand. This may also be the same for the industry sector which is expected to grow at 7.1 % by 2040, while commercial/services sector grows at 13.3 %. The agriculture sector however, followed the status quo of

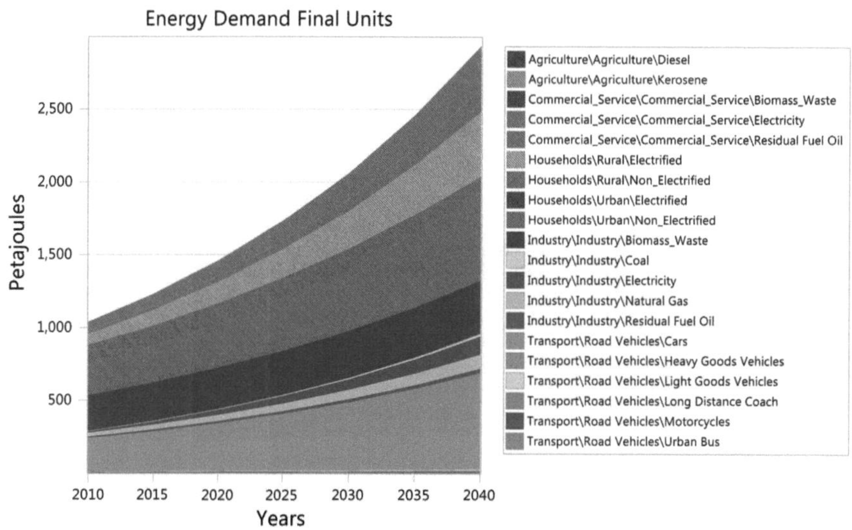

Fig. 5.10 LCM scenario energy consumption

the Ref scenario; this is due to the low energy demand of the sector as a result of lack of mechanized farming that might increase energy consumption from 2010 to 2040.

The total energy demand in the transport sector was 688.3 PJ, which is higher than that of Ref scenario (581 PJ). It should be taken into account that an increase in income and population was taken into consideration in the LCM scenario, which means more people will buy more cars as income increases and population size increases. The noticeable increase in energy demand in the transport sector was observed in cars with about 658.9 PJ in 2040 as against the energy demand of 557.2 PJ in the Ref scenario. The introduction of biofuels in the transportation fuel mix (see Table A.2) is bound to increase energy demand but may reduce GHG emission as it will later be discussed in subsequent sections.

Future Energy Supply (LCM)

The total energy supply (2,940.5 PJ) under the LCM scenario was lower than the ref scenario and this saw a shift in the fuel mix. This is shown in Fig. 5.11 and presented in Table D.3. Natural gas supply in the LCM scenario was reduced to 95.8 PJ as against the value in the Ref scenario (111.3 PJ), while domestic fuels which are usually in high demand (e.g. kerosene and gasoline) had a reduction in demand. The reduction in gasoline (371.6 PJ) in the LCM scenario was due to fuel switching in the transportation sector. However, the increased capacity of gasoline generators (14,000 MW) for off-grid electricity production did not cause an increase in gasoline fuel demand. This was also observed in the case of diesel fuel

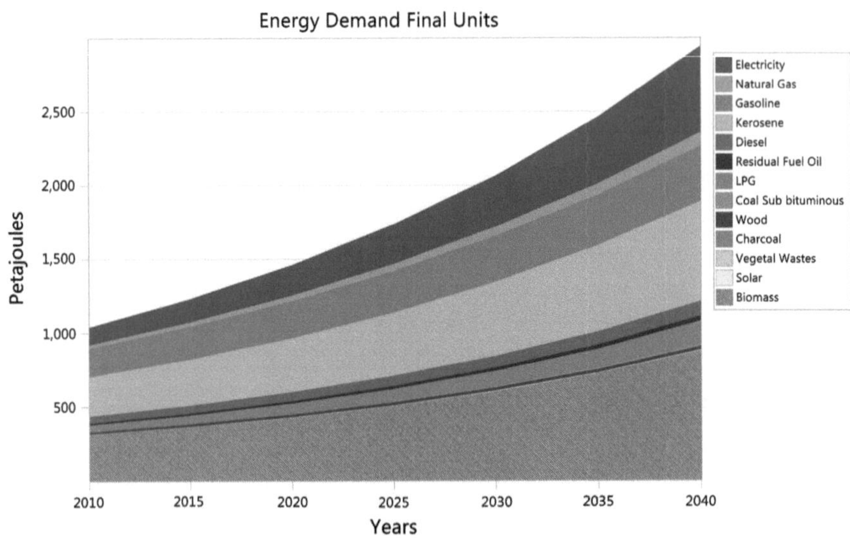

Fig. 5.11 LCM scenario energy supply

demand, with its demand accounting for 99.1 PJ in 2040 as compared to the higher demand in the Ref scenario (i.e.112 PJ). The increase in the demand for biofuels is expected to increase the demand for biomass from 312 PJ in 2010, to 880.1 PJ by 2040, which is higher than the Ref scenario.

5.1.2.2 Future Electricity Supply (LCM)

The installed capacity of 170,500 MW by 2040 under the LCM scenario is expected to generate about 171,800 MWh, and this is shown in Fig. 5.12 and presented in Table E.2. The bulk of electricity generation is expected to come from SCGT power plants (38,300 MWh) which make up about 31 % of the total on-grid electricity power plants (see Table A.4). CCGT power plants are expected to generate 28,900 MWh by 2040 from their installed capacity of 25,000 MW. The nuclear power plant capacity was increased from 4,000 MW in the Ref scenario to 5,000 MW in the LCM scenario and this will ensure the generation of 5,500 MWh of electricity by 2040 as against the 4,600 MWh generated from nuclear power in the Ref scenario.

Coal supercritical power plant with an installed capacity of 11,000 MW by 2040 will generate about 11,400 MWh of electricity. Coal supercritical power plants are by far more efficient and emit less GHGs as compared to the coal steam power plants. However, this scenario takes into consideration cheaper cost of energy technologies for the Nigerian power sector until 2040. The presence of renewables is low as compared to fossil fuel energy technologies. There was a little observable increase in electricity generation from hydropower (9,700 MWh).

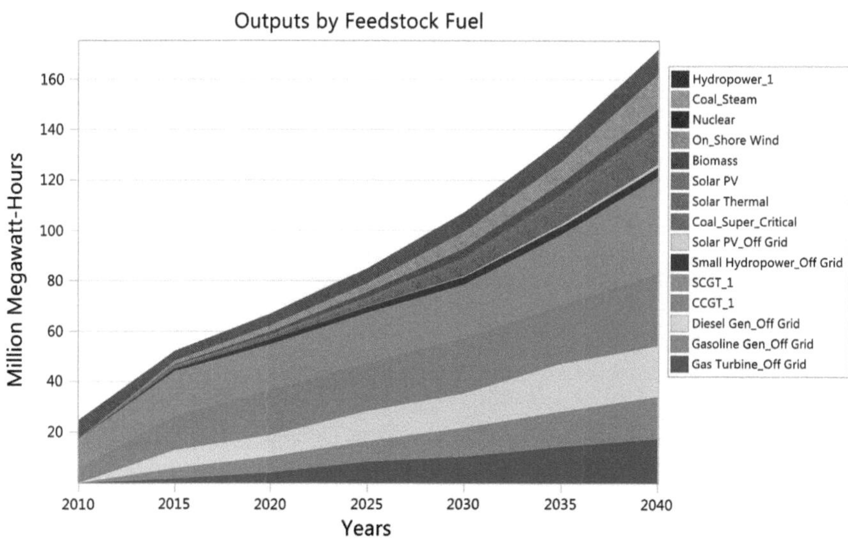

Fig. 5.12 LCM scenario electricity supply

Small hydropower plants, on the other hand, are expected to contribute about 3,500 MWh of electricity to off-grid areas in Nigeria by 2040 and will aid in the improvement of electricity access in Nigeria. Solar PV is expected to contribute to both on-grid and off-grid areas in Nigeria, with effect from 2020. On-grid electricity supply from solar PV in the on-grid areas will be 2,700 MWh, while off-grid areas will receive about 1,300 MWh of electricity from solar PV systems in off-grid areas. The LCM scenario also considered the increased expansion of off-grid fossil fuel plants such as diesel generators, gasoline generators, and gas turbine plants to generate about 20,200, 16,700, and 17,500 MWh respectively.

5.1.2.3 GHG Emission (LCM)

The main motive behind the development of the LCM scenario was to increase energy supply through the use of cheaper energy technologies, while moderately improving energy efficiency, and have a moderate reduction in GHGs. The total GHGs emission observed in the LCM scenario was 105.2 MMTCDE by 2040 as shown in Fig. 5.13 and presented in Table G.2. In this scenario, the households sector had reduced GHGs from energy consumption from 64.1 MMTCDE in the Ref scenario, to 58.4 MMTCDE by 2040 in the LCM scenario. The industry sector had GHGs amounted to 9.9 MMTCDE in 2040, from 2.8 MMTCDE in 2010. The transport and commercial/services sectors had a lower GHG emission rate of 2.3 and 0.1 MMTCDE in 2040 respectively. These values are lower than the Ref scenario, except for the agriculture sector which gave similar situation as the Ref scenario. The transport sector which saw the inclusion of biofuels in the fuel

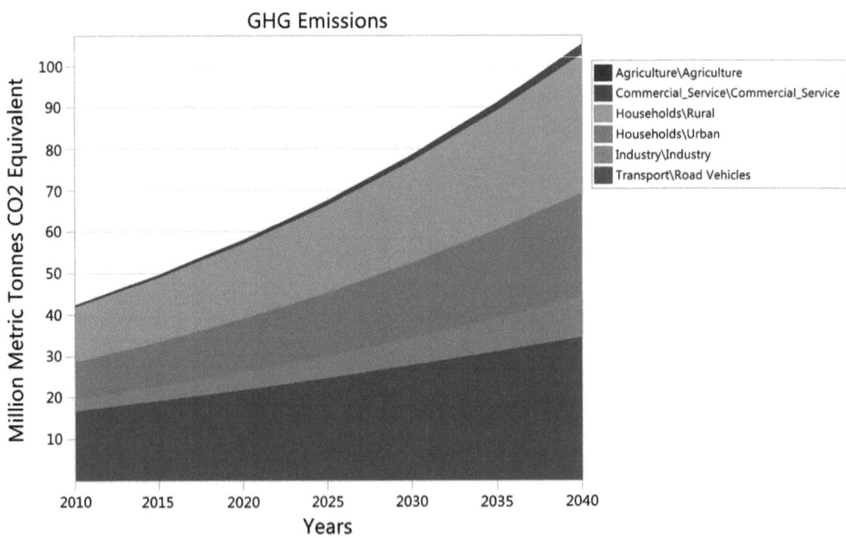

Fig. 5.13 LCM scenario GHG emission from energy demand

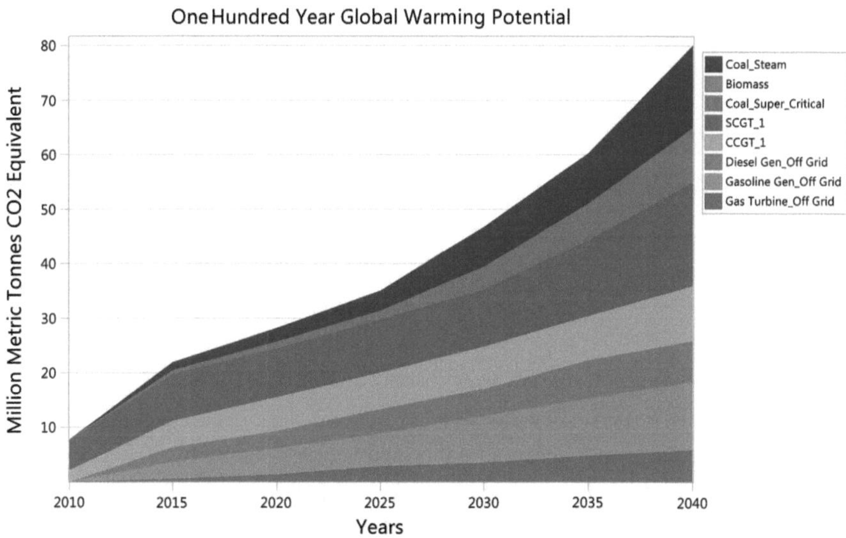

Fig. 5.14 LCM scenario GHG emission from electricity generation

mix, with an intention to reduce GHG emission, achieved a reduction to about 34.5 MMTCDE by 2040. This reduction is by far a boost to the transportation sector, as it provides a cleaner fuel and makes provisions for fuel options.

The GHGs from electricity generation on the other hand (Fig. 5.14 and Table G.6), had the total emission increased to 80.1 MMTCDE by 2040 from 7.8 MMTCDE in 2010. The reduced capacity of the SCGT plants in the LCM scenario as compared to the Ref scenario, led to the reduction in its GHGs to 19.2 MMTCDE. However, this will still contribute the highest GHG emissions when comparing to GHGs emission of other fossil fuel power plants in the electricity mix.

Diesel generators for off-grid electricity production contributed less amount of GHGs when compared to their counterpart energy technology, gasoline generators. The installed capacity of diesel generators for off-grid was 17,000 MW (about 30 % share of off-grid), while gasoline generators was 14,000 MW (about 24 % share of off-grid) and in comparison, diesel generators emitted less GHGs than gasoline generators. Coal supercritical power plant and coal steam power plant under the LCM scenario emitted 15.2 MMTCDE and 9.8 MMTCDE, respectively in 2040. This however, may be termed as high amount of GHGs from power generation, but the concept of cheaper energy technology and available fuels need to be taken into consideration in this scenario.

5.1.2.4 Energy Balance (LCM)

Under the LCM scenario, Nigeria's energy balance differs from the Ref scenario and this is shown using a Sankey diagram in Fig. 5.15 and presented in Table H.3.

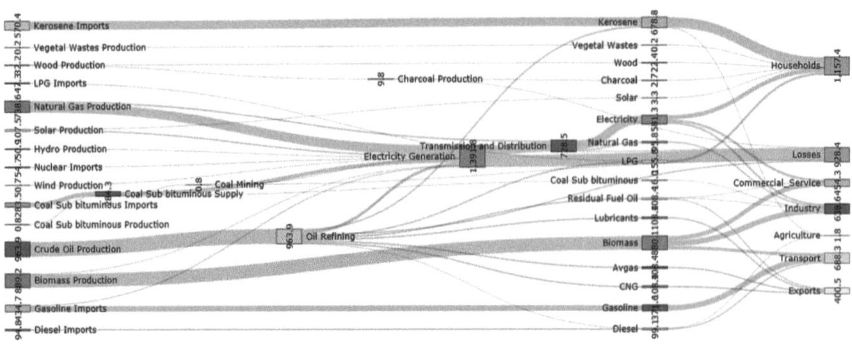

Fig. 5.15 Sankey diagram of Nigeria's energy balance in 2040 (LCM Scenario)

According to the diagram, Nigeria's total primary energy supply by 2040 is expected to amount to 3,868.9 PJ. The increased capacity of power plants under this scenario eliminated the need for importation of electricity, since the production of 618.4 PJ of electricity is able to meet sectorial electricity demand (i.e. households, industry, commercial/services).

Natural gas import by 2040 under the LCM scenario will not be required because its domestic production of 738.6 PJ is sufficient to meet demand for power plants sector (628.5 PJ) and industry sector (95.8 PJ). Unlike natural gas, petroleum products such as gasoline, kerosene, diesel, and LPG will continue to be imported to meet domestic demand up to 2040. Wood production of 32.2 PJ will meet the domestic demand for charcoal production (9.8 PJ) and firewood (or fuel wood) demand of 22.4 PJ.

As shown in the chart diagram of the energy balance in Fig. 5.16, the increased demand for LPG will require the importation of additional 47.3 PJ to complement the domestically produced 108.4 PJ to meet the demand of 155.8 PJ.

From Fig. 5.16, biomass produced from indigenous source[13] of about 795.4 PJ will meet its domestic demand of 426.8 PJ for industry and 368.4 PJ for commercial/services, while about 0.3 PJ will be consumed by the biomass power plants[14] in 2040. Wood in this scenario will be used to meet household demand (firewood) and for charcoal production, since there is an improvement in types of kilns used for charcoal production.[15] Other fuels such as CNG, Avgas, and lubricants are exported to the international markets since there were no domestic requirements[16] for the fuels.

[13]It is important to note that biomass production differs from wood and charcoal production, because when we say "wood", we mean firewood for domestic consumption and this is different from biomass used for electricity generation and industrial or commercial/services consumption.

[14]This is also shown in Fig. 5.16 as the secondary requirement.

[15]Please refer to Table A.3 to see the types of kilns used in the LCM scenario.

[16]The LCM scenario did not assume the use of these fuels in the development of the Nigeria LEAP model. However, these fuels are actually in high demand in Nigeria (except for CNG), but due to lack of available data for the consumption of the fuels, this scenario did not consider the fuels in the energy mix. In other scenarios, CNG will be considered in the energy mix.

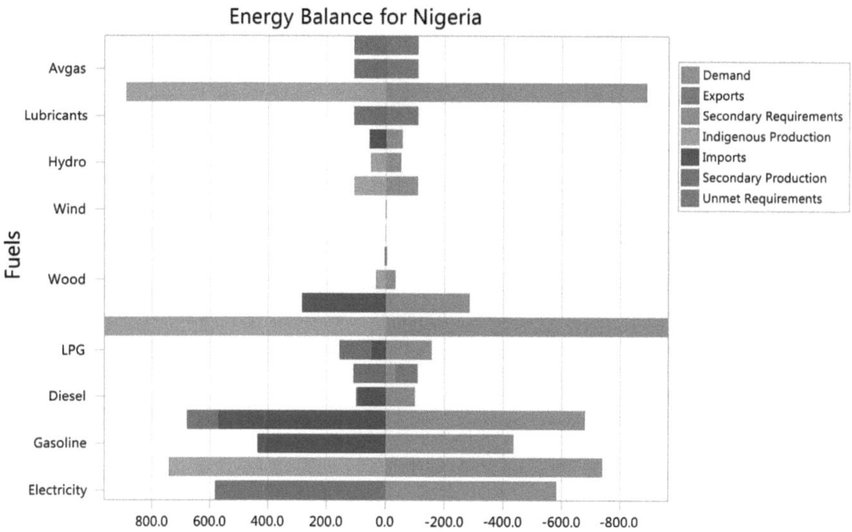

Fig. 5.16 Nigeria's energy balance in 2040 (LCM Scenario). (Order of fuels; CNG, Avgas, biomass, lubricant, nuclear, hydro, solar, wind, vegetal waste, charcoal, wood, coal sub-bituminous, crude oil, LPG, residual fuel oil, diesel, kerosene, gasoline, natural gas, electricity.)

5.1.3 The Low Carbon Advance Scenario (LCA)

The motivation for the development of this scenario is to move Nigeria towards a cleaner fossil fuel economy and aggressively reduce energy demand through policies and strategy. The results are presented as follows:

5.1.3.1 Future Energy Demand and Supply (LCA)

Future Energy Demand by Sector (LCA)

This scenario presents a much lower energy demand rate as compared to the last two scenarios (i.e. Ref and LCM). The total energy demand under the LCA scenario for the year 2040 is 2,488.3 PJ and this is by far the lowest recorded energy demand in energy-intensive sectors. Household sector's energy demand was 948.3 PJ in which the urban households accounted for 403.6 PJ, while the rural household's energy demand was 544.7 PJ. As in the case of the LCM scenario, the non-electrified urban households are assumed to be totally connected to the grid by 2040, while electricity access is improved in the rural areas.

Energy demand in the industry sector had a reduction of about 471.6 PJ by 2040, which was an improvement compared to the LCM and Ref scenario. Out of this reduction, biomass/waste demand was 132.1 PJ, while natural gas, residual fuel oil, coal and electricity was 117.9 PJ, 23.6 PJ, 9.4 PJ, and 188.7 PJ

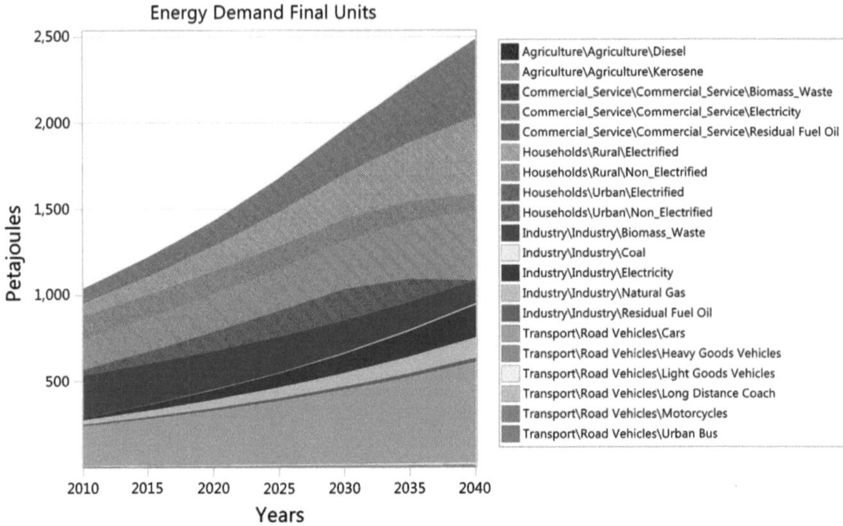

Fig. 5.17 LCA scenario energy demand

respectively. A similar situation was observed in the commercial/service sector which had an energy demand of 454.3 PJ. The energy demand in this sector was mostly for biomass/waste (234.9 PJ) and electricity (216.7 PJ), while a much lower requirement of 2.7 PJ came from residual fuel oil. The results of the energy demand under the LCA scenario are shown in Fig. 5.17 and presented in Table C.3.

Future Energy Supply (LCA)

The required energy supply expected to meet the energy demand under the LCA scenario by 2040 is 2,249.2 PJ. This is shown in Fig. 5.18 and Table D.4. From the results, the major form of energy that will be supplied by 2040 will be electricity which is about 689.1 PJ, followed by biomass and kerosene with 531.8 PJ and 525.5 PJ, respectively. This scenario revealed the increase in electricity supply and reduction in kerosene and biomass as compared to previous scenarios (i.e. Ref and LCM).

Gasoline supply was reduced to 158.7 PJ as compared to the LCM scenario which had its supply of gasoline to the value of 246 PJ, while natural gas supply increased to 121.5 PJ from 95.8 PJ in the LCM scenario by 2040. An increase in LPG fuel supply was observed to amount to 312.2 PJ and can be attributed to the introduction of LPG vehicles in the transport sector.

5.1.3.2 Future Electricity Supply (LCA)

In modeling the electricity supply under the LCA scenario, consideration was given to more efficient, low-carbon power plants such as CCGT, nuclear, coal

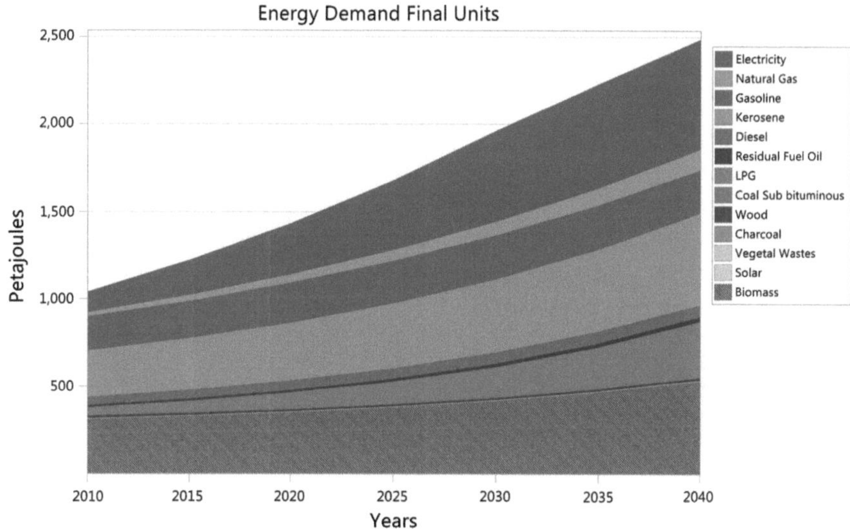

Fig. 5.18 LCA scenario energy supply

IGCC, and CFB power plants. The total electricity generated under this scenario amounted to 184,300 MWh from the installed capacity of 180,500 MW by 2040. In the electricity generation mix, CCGT accounts for 39,300 MWh of power generation from the installed capacity of 30,000 MW, while the less efficient SCGT plants for on-grid electricity supply will generate 14,900 MWh of electricity in 2040. Off-grid CCGT plants which had an increased capacity unlike the Ref and LCM scenarios, will generate about 14,900 MWh which is the same generation capacity as the on-grid SCGT.

The introduction of coal IGCC and CFB technology with an installed capacity of 13,000 and 14,000 MW respectively, is expected to generate 15,200 and 16,400 MWh respectively from coal energy resources. The fuel for the IGCC and CFB power plants (i.e. sub-bituminous coal) is expected to be sourced domestically, but this will depend on the future expansion plans for coal mining in Nigeria. Gas turbine power plants for off-grid electrification will be expanded to 11,000 MW capacity to be capable in the generation of 12,100 MWh by 2040. This will be supported by the SCGT power plants for off-grid power generation, because of the retirement of the gasoline and diesel generators before 2015.[17] The results for electricity generation under the LCA scenario are shown in Fig. 5.19 and presented in Table E.3.

Renewables in the electricity mix include biomass with an increased electricity supply of 8,600 MWh and this is due to the increased installed capacity of 7,500

[17]This is shown in Table A.6, and the intention was the quest for cleaner energy technologies of the LCA scenario, as well as the assumption that natural gas will become readily available for natural gas power plants in both off-grid and on-grid areas.

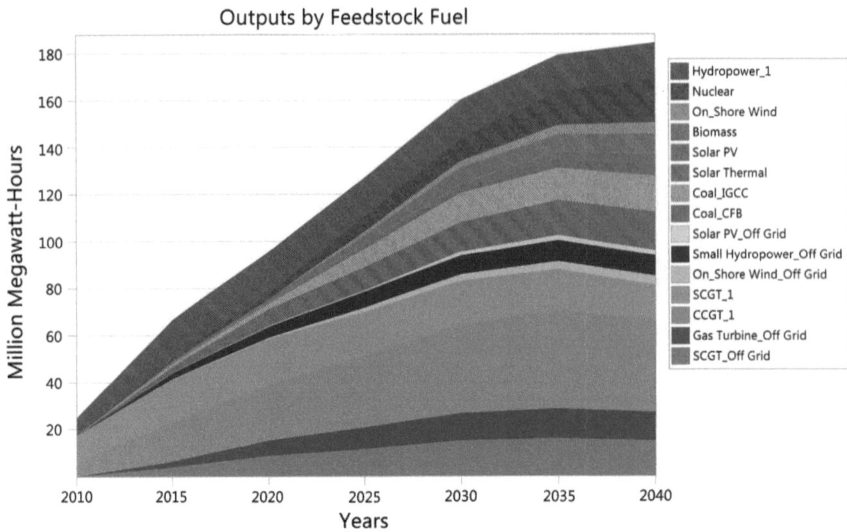

Fig. 5.19 LCA scenario electricity supply

MW in 2040. Large hydropower electricity generation was modeled to generate 15,400 MWh in 2040, small hydropower which was installed for off-grid generation produced 8,800 MWh of electricity. The increased capacity of small hydropower was intended to support the gas and SCGT turbine. Solar PV and thermal systems are expected to increase capacity from 2015 to 2040, and this will contribute 4,200 MWh for solar PV and 5,200 MWh for solar thermal. The solar PV and thermal systems were intended to come in the form of solar farms, which will contribute a significant amount of electricity to the grid. On-shore wind power will also have an increased capacity to be able to generate 4,700 MWh of electricity from its installed capacity of 8,000 MW by 2040 for on-shore, and 3,900 MWh for off-shore from an installed capacity of 6,500 MW.

5.1.3.3 GHG Emission (LCA)

The aim of achieving a low-carbon society with cleaner fuels and efficiency practice in the LCA scenario led to the reduced total GHG emission of 90.8 MMTCDE by 2040. The household sector witnessed the highest reduction in GHGs of 48.4 MMTCDE, in which the urban household emitted less GHGs as compared to the rural households. This is shown in Fig. 5.20 and Table G.3. The industry sector in this scenario emitted less GHGs of about 9.4 MMTCDE, as compared to the Ref (11.5 MMTCDE) and LCM (9.9 MMTCDE) scenario.

The commercial/services sector had reduced GHGs of 2.0 and 0.1 MMTCDE respectively in 2040. The reduction in GHG emission in the commercial/services sector was due to fuel switching whereby the energy demand for biomass

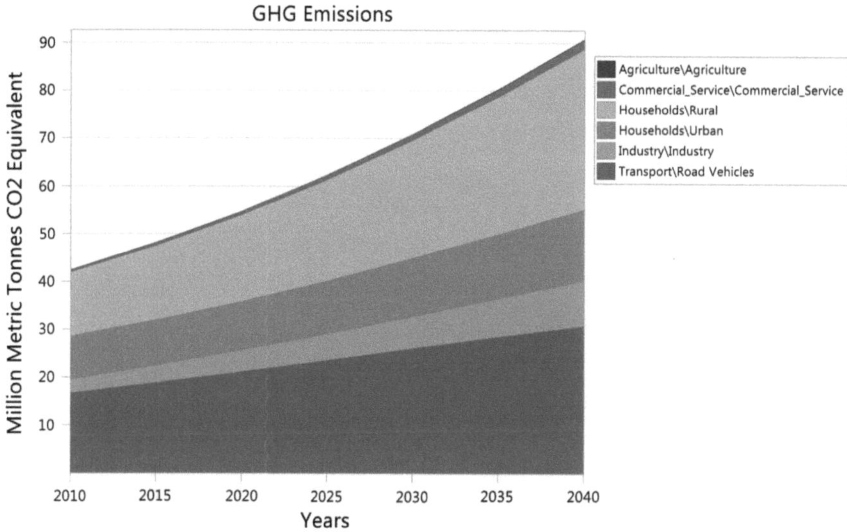

Fig. 5.20 LCA scenario GHG emission from energy consumption

was reduced, while electricity supply was increased. The last sector which is the transport sector had a decrease in GHGs emission to 30.8 MMTCDE in 2040. The reduction was much lower, compared to the LCM (34.5 MMTCDE) and Ref (40.7 MMTCDE) scenarios.

On electricity generation, the GHG emission increased from 7.8 MMTCDE in 2010, to 61.2 MMTCDE by 2040 and this can be observed to be very low when compared to the previous scenarios. The increased generating capacity of coal plants (IGCC and CFB) unlike the previous scenarios, led to the combined GHGs of 28.5 MMTCDE for both coal power plants. These results are shown in Fig. 5.21 and presented in Table G.7. The emission level recorded for SCGT plants for on-grid was 7.5 MMTCDE by 2040, and this was the same amount for the SCGT off-grid power plants. The gas turbine power plants for off-grid were observed to emit less GHGs of about 4.1 MMTCDE when compared to SCGT power plants. When the capacity is compared, it is about a 1000 MW difference between the two power plants.[18]

5.1.3.4 Energy Balance

The energy balance under the LCA scenario is bound to differ from the previous scenarios and this is shown in Fig. 5.22 and presented in Table H.4. The total

[18]This situation can be observed in Table A.6, where the installed capacity for the off-grid SCGT was 12,000 MW, while the off-grid gas turbine was 11,000 MW. One would expect close GHGs or even less GHGs from the SCGT power plant. This will however, be explained in the discussion section of this chapter.

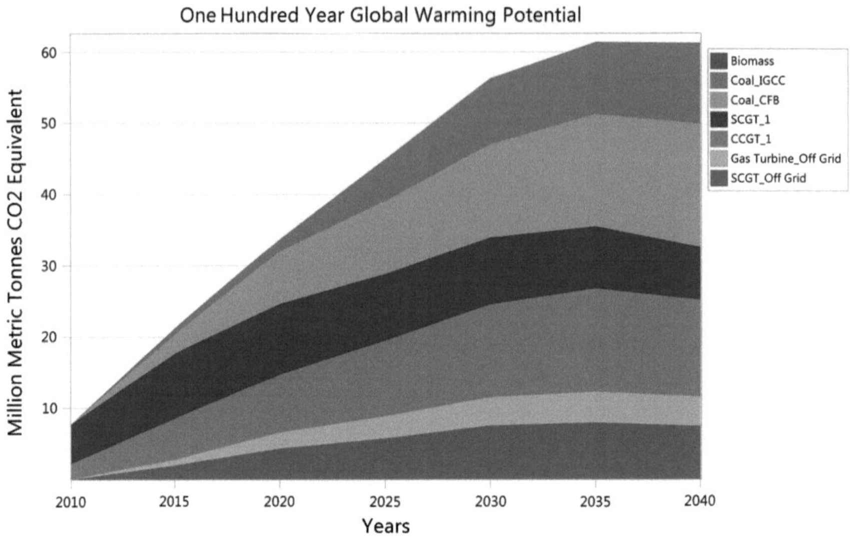

Fig. 5.21 LCA scenario GHG emission from electricity generation

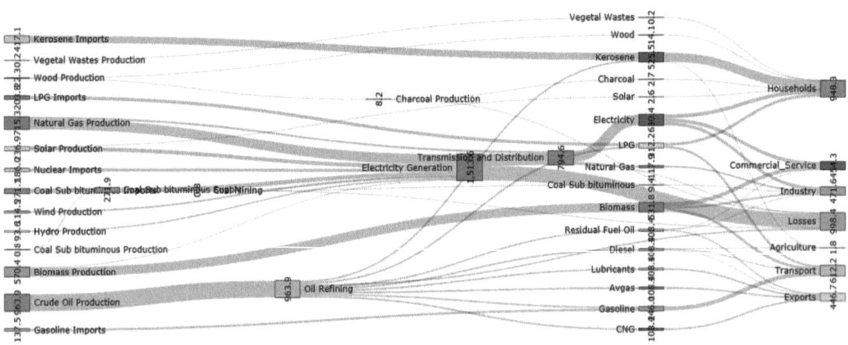

Fig. 5.22 Sankey diagram of Nigeria's energy balance in 2040 (LCA Scenario)

primary energy supply under this scenario is 3,486.7 PJ and this was at a reduced level unlike the LCM and Ref scenarios.

Electricity generation in this scenario in 2040 was 663.5 PJ and this is sufficient to meet the sectorial electricity demand of 630.4 PJ, even after the T&D losses (i.e. 33.2 PJ). Although electricity demand for households was 225 PJ, which was lower than the LCM and Ref scenarios households demand, the industry and commercial/services sector had higher demand. The increase in electricity demand in industry and commercial/services sectors is due to the fuel/energy switching.[19] Also, the reduced T&D electricity demand was due to the improvement in T&D losses which was reduced to 5 % by 2040.

Natural gas produced from domestic refineries was sufficient to meet the demand for natural gas in the industry sector and for electricity generation, while loss due to T&D was 13.1 PJ. Since gasoline generators were retired before 2015 under this scenario, the demand for gasoline by the transport sector increased and amounted to 246 PJ. This demand was however, sourced from both secondary production[20] and imports as can be observed in Fig. 5.23.

Kerosene also followed suit as gasoline with the importation of 417.1 PJ of kerosene to complement the domestic production of 108.4 PJ, which will be able to meet the demand of 525.5 PJ in the household and agriculture sector. The demand for diesel fuel on the other hand, will be met by productions from domestic oil refineries. This is because the power sector will have its diesel plants retired as far back as 2015, while demand in the transport sector will shrink to 67.7 PJ as compared to the previous scenarios. However, about 39.3 PJ of diesel fuel will be exported to the international market since the fuel will be produced in excess.[21]

The increased use of LPG in the transport sector will increase its demand to 133.7 PJ by 2040, while 178.6 PJ will be required to meet household demand for cooking and lighting. The demand for LPG will be met through importation (203.8 PJ) and secondary production of 108.4 PJ (see Figs. 5.22 and 5.23). The coal mining capacity was insufficient to meet the demand for sub-bituminous coal and as such, required the import of about 271.1 PJ to meet both the secondary requirement for electricity generation (262.3 PJ) and industry demand of 9.4 PJ.

The increase in nuclear power plant capacity in the future (i.e. 2040) to 15,000 MW will require the importation of 186 PJ of nuclear fuel (uranium). Biomass production from domestic sources will amount to 570.4 PJ which will meet biomass power plant demand of 38.6 PJ for its 7,500 MW installation. Other demand for biomass will come from industry with the demand of 132.1 PJ, while commercial/services will require 234.9 PJ to meet its energy demand. Biofuel which is produced from biomass will have an energy demand of 164.9 PJ in the transportation sector. In all, out of the total primary energy supply of 3,486.7 PJ in 2040, 2,488.3 PJ will be required to meet sectorial energy demand, while about 998.4 PJ will be required for the transformation of individual sector.

[19]In this case, the industries and commercial/service sectors used more electricity and less forms of other fuel/energy sources.

[20]This means production from domestic refineries which was considered to have the base year capacity continues until 2040 with no expansion plans.

[21]This however, may still be sold to private diesel operators and to the waterways transporters (ships and boats), or may be exported to neighboring countries.

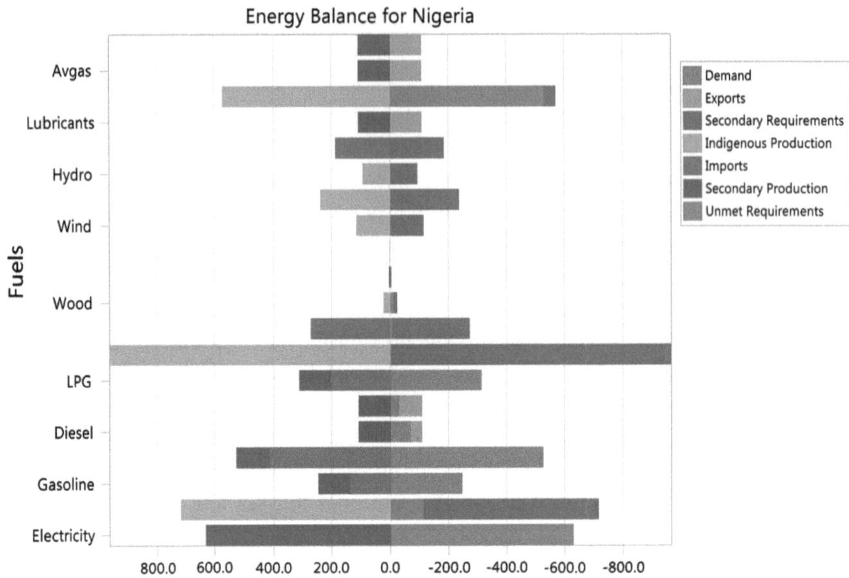

Fig. 5.23 Nigeria's energy balance in 2040 (LCA Scenario). (Order of fuels; CNG, Avgas, biomass, lubricant, nuclear, hydro, solar, wind, vegetal waste, charcoal, wood, coal sub-bituminous, crude oil, LPG, residual fuel oil, diesel, kerosene, gasoline, natural gas, electricity.)

5.1.4 The Green Optimistic Scenario (GO)

This scenario was developed in view of Nigeria attaining a low-carbon-green growth economy which will contribute in the mitigation of climate change, reduce energy poverty, and ensure energy sustainability. The results of G.O scenario are presented as follows:

5.1.4.1 Future Energy Demand and Supply (GO)

Future Energy Demand by Sector (GO)

The total energy demand under the GO scenario was observed to be 2,249.2 PJ which accounted for the lowest energy demand among the four scenarios proposed in this study. The highest energy demanding among the five sectors was the household sector with 836 PJ of energy demand by 2040. Out of this, the rural households had the largest share of energy demand with 552.8 PJ, while the urban households' energy demand was 283.1 PJ. The results are shown in Fig. 5.24 and Table C.4. From the results, it can be observed that rural energy demand increased in the GO scenario when compared to the previous scenarios. However, energy demand in the urban sector was reduced unlike the situation observed in the previous scenarios.

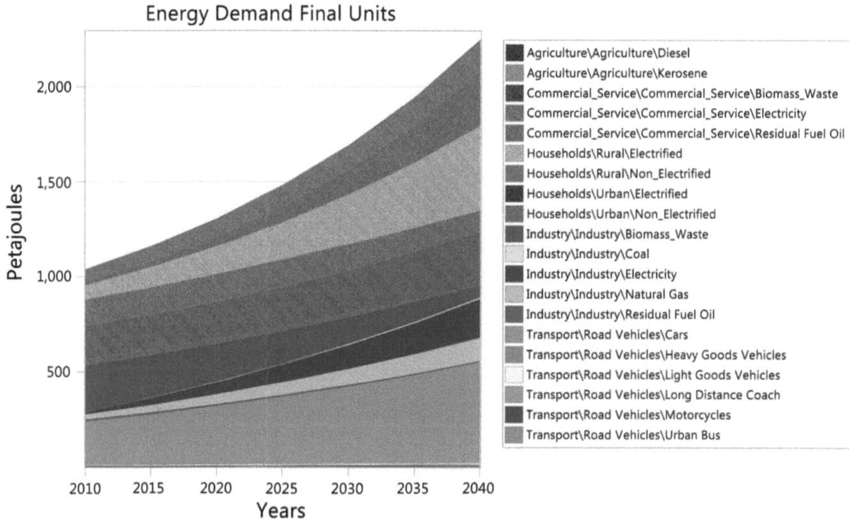

Fig. 5.24 GO scenario energy demand

The energy demand in the industry sector is expected to amount to 404.9 PJ by 2040 from its 2010 values of 297 PJ in 2010. This was observed to be a decrease as compared to the LCA, LCM, and Ref scenarios. Along with the reduction in energy demand, biomass which had the highest fuel demand in the base year and in the Ref and LCM scenario, was reduced to 64.8 PJ by 2040. Natural gas demand on the other hand, gradually increased in its share of the fuel mix to 121.5 PJ from 25.4 PJ in 2010. The demand for other fuels such as coal and residual fuel was observed to have a near value of 8.1 PJ each. This scenario assumes an increase in electricity demand in the industry sector due to the introduction of electric arc blast furnace.[22]

The commercial/service sector's energy demand increased from 84 PJ in 2010 to 454.3 PJ in 2040. These values are same for the total energy demand for the LCM and LCA scenarios, but the difference observed was in the energy mix. The difference was observed in the increased share of electricity demand which was 289.4 PJ by 2040. The energy demand for the agriculture sector was the same as the previous scenarios. In the transport sector, considerable reduction in energy demand was observed and the total energy demand for the sector was 552.2 PJ in 2040.

The energy demand by motorcycles for both private and commercial transport option was 5.6 PJ, while cars was 528.4 PJ in 2040. Besides the inclusion of various fuel types to create fuel options, the energy demand was still low as compared to other scenarios. Light- and heavy-duty vehicles had their demand for energy increase from their 2010 values, to 5.1 PJ and 0.9 PJ respectively in 2040. Energy demand for urban bus and long distance coach was 4.2 PJ and 8 PJ respectively.

[22]This was fully discussed in the scenario development section (GO scenario) in Chap. 4.

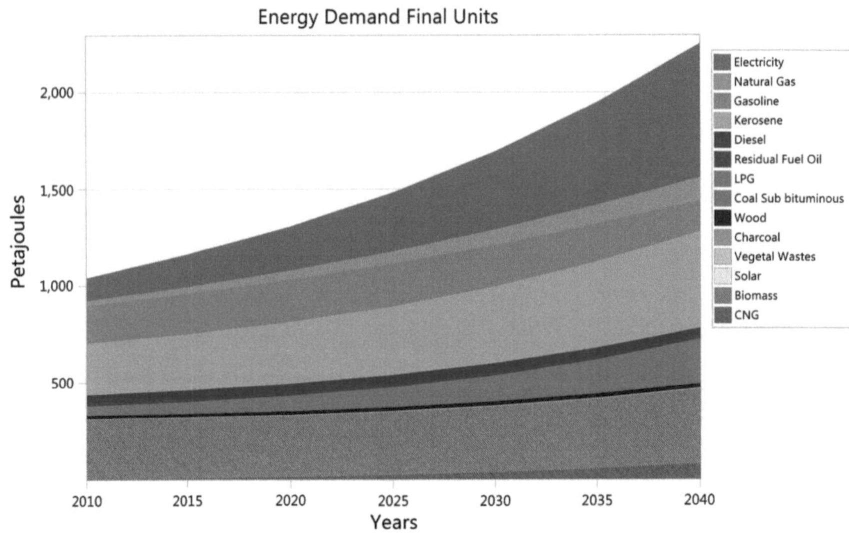

Fig. 5.25 GO scenario energy supply

Future Energy Supply (GO)

The energy supply by fuel required to meet the energy demand in the sectors analyzed under the GO scenario is shown in Fig. 5.25 and presented in Table D.5. The total energy supplied by 2040 is 2,249.2 PJ and the energy type with the largest share was electricity with a supply rate of 689.1 PJ. There is an increase in electricity supply rate when compared to the LCA (630.4 PJ), LCM (581.3 PJ), and Ref (582.5 PJ) scenarios. Natural gas was observed to have an increased value of 121.5 PJ in 2040. This is due to increased use of natural gas in the transformation sector. The reduction in the share of gasoline vehicles and the retirement of gasoline power plants before 2015 are responsible for reduction in its supply to 158.7 PJ in 2040, from 193.8 PJ in 2010.

Kerosene supply in the GO scenario was estimated to be 499.4 PJ by 2040. This is the lowest supply value when compared to previous scenarios. The supply of diesel fuel was found decreasing from year 2010 to 2040, and this was observed to occur between the years 2030–2035.[23] As the demand for LPG grew due to its increased use in the transport and household sector, the supply rate increased to about 223.3 PJ by 2040. Wood supply increased from 15.4 PJ in 2010 to 22.2 PJ by 2040. Biomass supply slowly increased from 312 PJ in 2010 to 386.7 PJ in 2040. As CNG was launched in 2015, its supply increased to 80 PJ by 2040.

[23]This coincides with the reduction in diesel cars around that same period as alternative fuels were introduced into the transportation sector.

5.1.4.2 Future Electricity Supply (GO)

Considering the advancement to a low-carbon-green growth in Nigeria, the GO scenario presented an increased share of renewables in the electricity mix. With the total installed capacity of 180,000 MW, electricity generation was 199,400 MWh by 2040. In the mix, large hydropower plants generated 20,100 MWh, from its installed capacity of 15,500 MW in 2040 and this contributes 11 % of the on-grid electricity power plants. The installed capacity of small hydropower for off-grid electricity production was 8,000 MW and this generated 11,000 MWh.

The GO scenario assumes the retirement of all fossil fuel power plants for off-grid electricity generation, except the SCGT power plants which will generate 5,800 MWh of electricity. However, to complement the retired fossil fuel power plants, biomass was added to improve electricity supply with the production of 8,100 MWh. Biomass was also installed for on-grid electricity generation and this produced 10,700 MWh of electricity. The results electricity generation under the GO scenario is shown in Fig. 5.26 and presented in Table E.4.

In order to expand the share of renewables and also include those not currently installed in the base year and also considered in previous scenarios, geothermal and off-shore wind power were introduced into the energy mix. These energy technologies were introduced in 2015 with an installed capacity of 1,000 MW for geothermal and 100 MW for off-shore wind power. The geothermal power plant will have its capacity increased up to 2040 and will generate 12,400 MWh, while the off-shore wind power will generate about 6,600 MWh from its increased capacity of 9,000 MW by 2040.

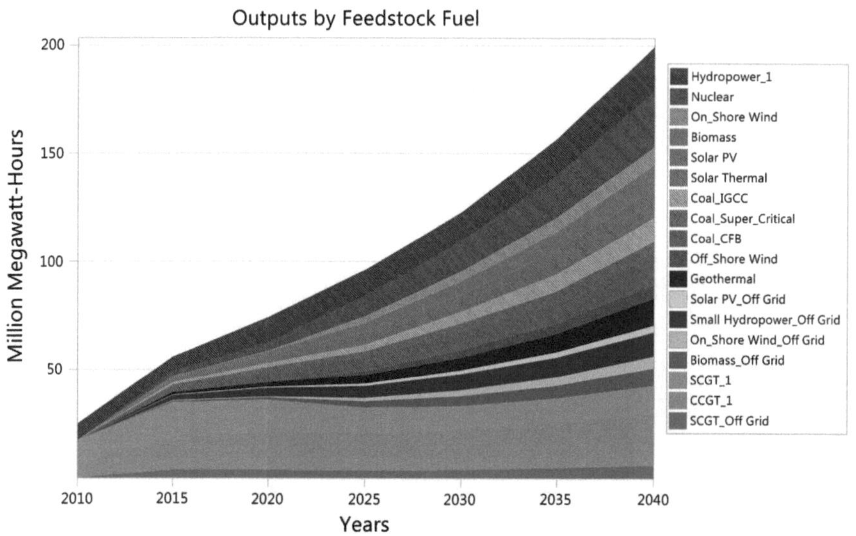

Fig. 5.26 GO scenario electricity supply

Solar PV systems which will be on a large scale farm had an installed capacity of 17,000 MW for on-grid and 9,000 MW for off-grid electricity generation. On-grid electricity generation from solar PV systems will be up to 6,100 MWh, while off-grid electricity generation will be 3,200 MWh by 2040. Nuclear power plants had an increased expansion of 18,000 MW by 2040 and this generated 26,200 MWh of electricity. This capacity is observed to be the largest installed capacity when compared to previous scenarios.

Meanwhile, the fossil fuel power plants in the GO scenario include gas fuel and coal fuel power plants. In a bid to invest in the fossil fuel power plants with higher efficiency, CCGT power plants had an increased capacity when compared to SCGT power plants. Electricity generation from CCGT power plants amounted to 24,500 MWh, while 12,400 MWh was generated from SCGT plants. For coal power plants, a mixture of three energy technologies which include IGCC, super critical, and CFB generated 11,000, 6,900, and 13,100 MWh respectively by 2040.

5.1.4.3 GHG Emission (GO)

Considering the fact that the GO scenario emphasizes on the expansion of renewables in order to achieve a low-carbon-green growth, a much lower GHG emission is expected from this scenario. The total GHG emission for energy demand under this scenario was 78.9 MMTCDE in 2040, and this is by far the lowest among the four scenarios. The result is shown in Fig. 5.27 and presented in Table G.4. The household GHG emission was 42.9 MMTCDE, which was a reduction from the LCA (48.4 MMTCDE), LCM (58.4 MMTCDE) and Ref (64.1 MMTCDE). The

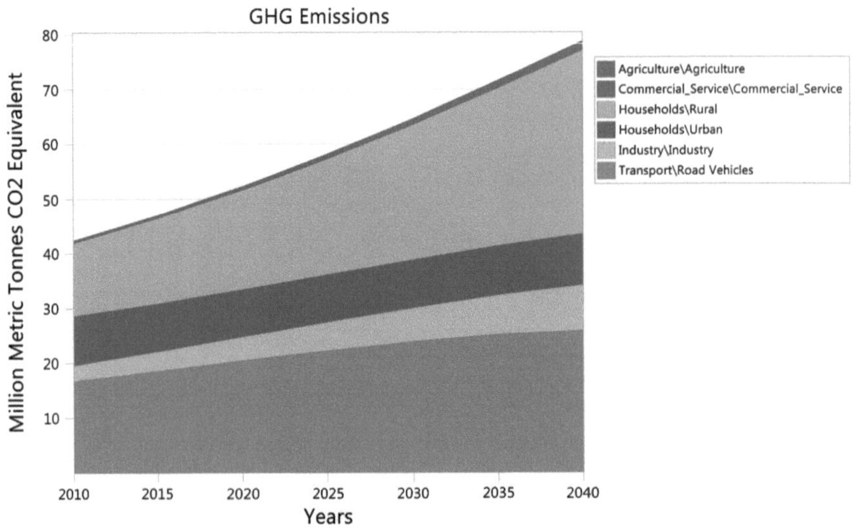

Fig. 5.27 GO scenario GHG emission from energy consumption

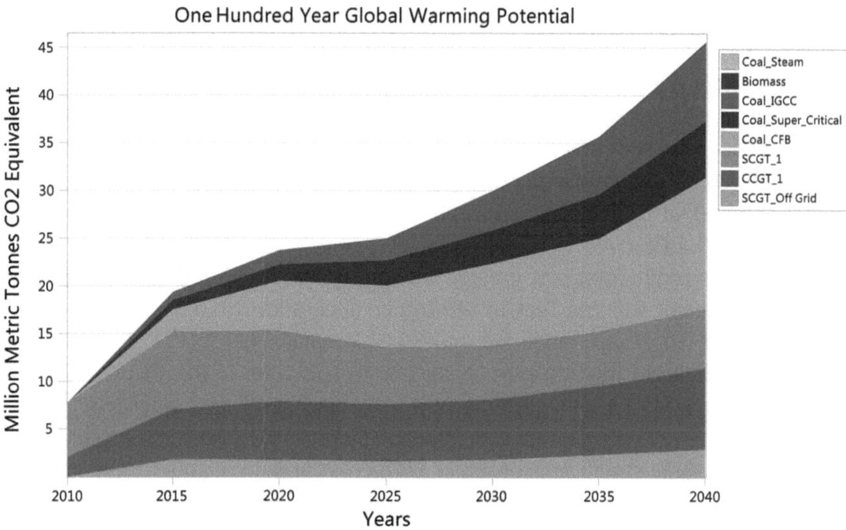

Fig. 5.28 GO scenario GHG emission from electricity generation

improvement in efficiency and fuel switching in the industry led to the decrease in GHGs from the sector to 8.3 MMTCDE by 2040. The commercial/service sector recorded a reduction of about 1.6 MMTCDE by 2040, while the transport sector had a reduced GHG emission of about 26 MMTCDE.

The result for the GHG emission from electricity generation in the GO scenario is shown in Fig. 5.28 and presented in Table G.8. The fossil fuel power plants in the GO scenario emitted a total of 45.6 MMTCDE in 2040 from its 2010 emission rate of 7.8 MMTCDE. Much of these emissions were from coal CFB power plants with an emission rate of 13.7 MMTCDE from its generation capacity of 9,500 MW.

Coal IGCC power plants emitted 8.2 MMTCDE in 2040, and this was similar to the emission recorded from CCGT power plants of installed capacity of 16,000 MW. An observation from these two power plants (i.e. IGCC and CCGT) is that even with the large capacity of 16,000 MW for CCGT power plants, the emission rate was lower when compared to the coal IGCC power plant of 8,000 MW.[24] Biomass power plant had the lowest emission rate of 0.1 MMTCDE from its installed capacity of 8,000 MW in 2040.

5.1.4.4 Energy Balance (GO)

The energy balance for Nigeria in 2040 under the GO scenario presents a case where by the presence of renewables in the energy mix is of an increased value

[24]This will be explained in a more elaborate manner in the discussion section of this chapter.

when compared to the previous scenarios. The total primary energy supply for this scenario was 3,507.5 PJ by 2040 and this was less than the Ref and LCM scenario, but a little over the LCA scenario. Electricity generation in energy unit was 717.9 PJ, and this had a loss of 28.7 PJ due to T&D losses.[25] The total transmitted electricity to meet the sectorial electricity demand was 689.1 PJ. The demand sector includes households with an electricity demand of 197.3 PJ, industry with an electricity demand of 202.4 PJ, and commercial/service sector with an electricity demand of 289.4 PJ.

Natural gas production was greatly reduced to 445.7 PJ, and this is due to the decreased demand for this fuel in electricity generation (316.5 PJ), lower T&D losses[26] (–7.8 PJ), and moderate industry demand (121.5 PJ). Gasoline has decreased demand in the transport sector due to reduction in the share of gasoline vehicles and the absence of gasoline power plants for off-grid electricity generation. Hence, there, will be reduced import of only 50.2 PJ of the product from overseas. The same situation is observed for kerosene with quantity to be imported reduced to 391 PJ. This quantity has to meet household domestic demand of 499.1 PJ and 0.4 PJ to be channeled to agriculture sector. The reduction in demand for diesel in the transport sector will experience increase in its export of 67.1 PJ from Nigeria's domestic oil refineries by 2040.

The household sector energy demand for LPG under the GO scenario was 112.6 PJ, while its use for transportation was 110.8 PJ, which was a significant reduction when compared to previous scenario. This led to its reduced import to 114.9 PJ from overseas in order to increase the domestic refineries production of LPG (108.4 PJ). The inclusion of the three types of coal power plants with various rates of energy conversion efficiencies will require coal consumption of 263.5 PJ, while the industry sector will demand 8.1 PJ of coal for its energy use.

On renewables,[27] wind energy of about 240.4 PJ will be used for power generation in both off-shore and on-shore turbines by 2040. Solar PV and thermal systems will utilize 341.1 PJ of solar radiation for electricity supply (339.1 PJ) and domestic solar installation of 1.9 PJ. Nigeria's large and small hydropower in both on-grid and off-grid areas will require about 120 PJ of hydro. The energy balance for 2040 under the GO scenario is shown in Fig. 5.29 and presented in Table H.5.

Geothermal power plant which was considered to increase power supply under the GO scenario will require 212.2 PJ of geothermal energy to generate 12,400 MWh by 2040 from its 8,500 MW installed capacity.

---

[25]This accounts for the lowest rate of T&D losses in the power sector which translate to 4 % T&D loss.

[26]As described in the GO scenario development in Chap. 4, the loss in natural gas T&D was reduced to 6 % by 2040 from its 2010 value.

[27]Note that the supply of energy for renewables is not to be seen as a conventional energy source since renewable energy is non-depleting. The primary energy supply from renewables shown in the energy balance is for the modeling purpose alone. Therefore, the figures listed in the energy balance are the amount of energy supplied to meet the demand of renewables (e.g. solar radiation, hydro, wind speed, etc.).

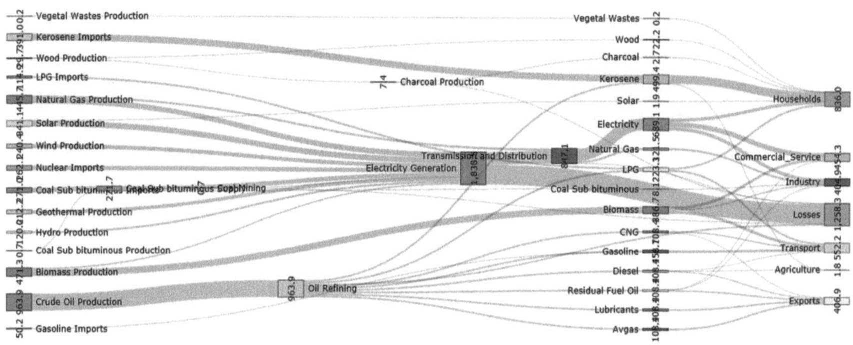

Fig. 5.29 Sankey diagram of Nigeria's energy balance in 2040 (GO scenario)

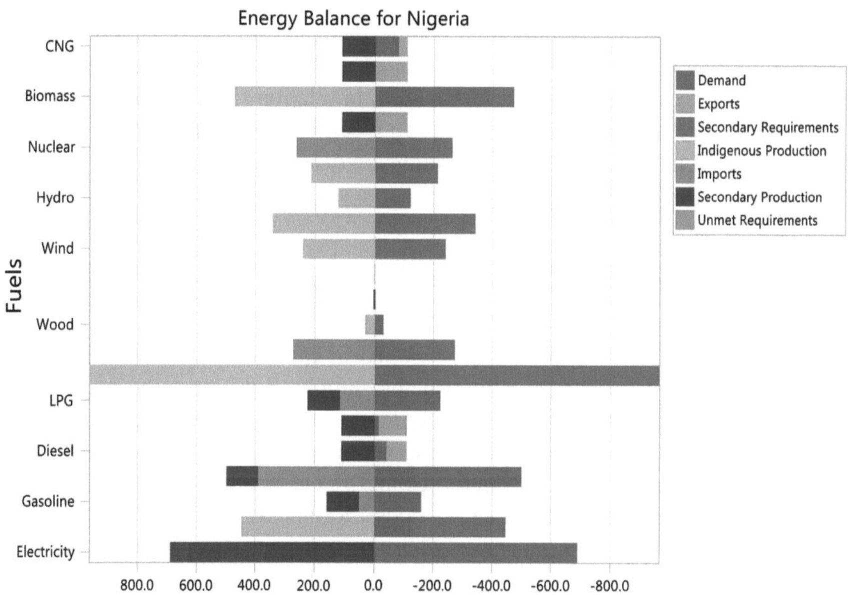

Fig. 5.30 Nigeria's energy balance in 2040 (LCA scenario). (Order of fuels; CNG, Avgas, biomass, lubricant, nuclear, geothermal, hydro, solar, wind, vegetal waste, charcoal, wood, coal sub-bituminous, crude oil, LPG, residual fuel oil, diesel, kerosene, gasoline, natural gas, electricity.)

The capacity expansion of 18,000 MW nuclear power plants by 2040 to generate 26,000 MWh of electricity will require the importation of 262.1 PJ of uranium fuel. This is also shown as secondary requirements in Fig. 5.30. It can also be observed from Fig. 5.30 that the indigenous production of biomass (471.3 PJ) will be used for domestic demand of 386.7 PJ and secondary requirement of 84.6 PJ by 2040.

Avgas and lubricants are exported to international markets or are used domestically but not considered in the Nigerian LEAP model. In the GO scenario, CNG vehicles were promoted and this reduced the amount of its export from domestic refineries (28.4 PJ), since its demand in the transport sector increased to 80 PJ by 2040. In conclusion, out of the total primary energy supply of 3,507.5 PJ in the GO scenario, 1,258.3 PJ was used in the transformation sector, while 2,249.2 PJ was used in the demand sector.

5.2 Discussions

This section analyzes the scenarios from a policy-based prospective in order to fully explore the effects of the strategic policies implemented in each scenario, while interrelating them together. In other words, this section examines the impact of policies on the energy supply mix, energy demand, electricity supply mix and GHG emissions in whole scenarios developed and the primary resource requirements. Also, the cost-benefits of each scenario in relation to the government based scenario (i.e. Ref scenario) and the least cost electricity generation from the power plants proposed in the scenarios in this study are discussed.

5.2.1 Meeting Energy Demand with Supply

Energy policies and strategies being developed in each scenario were intended to balance the demand for energy with available supply. But before the provision of energy to the demand sectors, measures or strategies have to be taken in ensuring that energy is efficiently utilized. The household sector is a case to begin with in the discussion of energy savings in Nigeria. Going by the Ref scenario or business-as-usual case, household energy demand is bound to increase to 1,267.6 PJ from the base year value of 418.5 PJ. However, energy efficiency polices such as the introduction of CFL bulbs and replacement of inefficient refrigerators in the LCM scenario will contribute to the reduction of energy intensities to 1,157.4 PJ. This means an energy saving of 110.2 PJ was achieved through this policy strategy alone, notwithstanding the increase in population, household and income growth rate.

The more aggressive steps taken in the LCA and GO scenario proved that with more energy efficiency strategies in place, more energy could be saved in Nigeria's household. This was realized through the introduction of LED bulbs for lighting with an intention of phasing-out incandescent bulbs in the LCA scenario. This strategic policy also came with the increase in LPG share for cooking and lighting[28] to 20 %, while inefficient air conditioning units and refrigerators were

[28]It should be taken into account that other fuels such as kerosene, firewood, charcoal, etc. also had their share reduced as shown in the energy supply results in the Appendix section.

replaced with efficient units. This increase in energy efficiency policy ensured the energy savings of 319.3 PJ from the LCA scenario.

Mixing LED and CFL with share of 40 and 60 % respectively by 2040 was the main focus of the energy efficiency policy in the GO scenario for lighting.[29] Besides the CFL and LED bulbs, the GO scenario ensured reduction in the share of fuels such as kerosene, firewood, and charcoal by encouraging LPG fuels and electricity instead. Solar thermal cookers, PV systems, efficient firewood and charcoal stoves were provided in the GO scenario. All these led to energy savings of 431.6 PJ by 2040.[30]

The same situation that occurred in the households sector was also observed in the commercial/service sector which had its electric light bulbs changed according to the alternative scenarios. In all, energy saving for the sector in the alternative scenario was 28.2 PJ, while the change in the energy mix in the sector also ensured that the energy supply was sufficient in reducing dependence on one fuel source.[31] Agriculture sector was not affected by policy changes due to lack of mechanized farming system associated with Nigerian farmers for the last two decades. However, the energy-intensive sector presents a reduction potential in energy demand due to policy interventions such as energy efficiency, technology upgrade, and fuel switching. The total energy demand in the four scenarios is shown in Fig. 5.31.

Although a significant level of energy efficiency policies was not common in the LCM scenario, the LCA goes a step further in ensuring the improvement in energy efficiency in industrial equipment which will reduce energy demand. Energy savings achieved due to the energy efficiency policy in the industry sector include about 103.6 PJ in the LCM scenario, 270.6 PJ in the LCA scenario, and 337.3 PJ in the GO scenario. In addition to the policies observed in the LCM and LCA scenarios, the GO scenario introduced a policy recommending the technology change of blast furnace from conventional blast furnace to electric-arc blast furnace. This technology change will in effect, reduce fossil fuel use in the industrial sector, while increase in electricity demand will be observed.[32]

In the transport sector there is a considerable amount of policy effect across the four scenarios. From the Ref scenario which had the demand for energy rising to 581 PJ by 2040, the impact of various strategies such as the provision of fuel options and efficiency improvement had a significant impact on energy savings. The introduction of biodiesel for motorcycles and biofuels for vehicles in the LCM scenario increased energy demand. This is a result of additional fuel options

[29]This ensured the complete phasing-out of incandescent bulbs before 2040.

[30]These were also some of the potentials discussed in Akinbami and Lawal (2009) where they estimated the future trend of electricity demand in Nigeria using the Model for Analysis of Demand for Energy (MADE-II). The study employed the useful energy demand concept and found considerable energy savings potentials in the residential and commercial sectors in Nigeria.

[31]This can be observed in the fuel switching from reduction of biomass to the increase in electricity supply.

[32]This is shown in Table C.4 for energy demand in the industry sector.

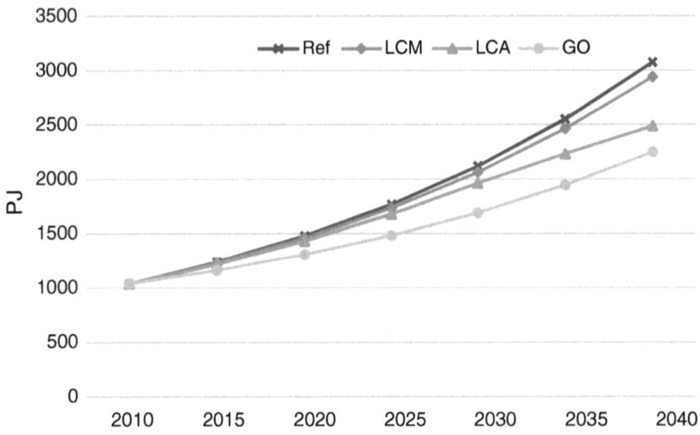

Fig. 5.31 Energy demand in the four scenarios

and increase in income.[33] The LCA scenario made a shift from the LCM scenario through the inclusion of LPG fuel in the transport sector. By 2040, the mix is expected to change as developed in Chap. 4. This will result in LPG competing with biofuel/biodiesel.

Unlike the LCM scenario, energy demand was much lower in the LCA scenario and this is due to fuel economy policy introduced. A more successful policy impact was observed for the transportation sector in the GO scenario. This feat is achieved through the inclusion of policies such as improved fuel economy for all vehicles, especially cars, the introduction of CNG, LPG, biofuels/biodiesel, gasoline and diesel into the transport mix; diesel was phased-out or reduced in some vehicle types such as cars which had no diesel engines by 2040, and long-distance coach and heavy-duty vehicles which had their shares reduced from the base year. The reduced energy demand due to the abovementioned policies resulted in energy savings of about 28.8 PJ for all vehicles in the transport sector under the GO scenario by 2040.

Energy demand among the alternative scenarios has been discussed in the previous section; the energy supply side will be the focus of this section. In meeting the demand for electricity in the three sectors (i.e. households, industry, and commercial/service), about 582 PJ of electricity was supplied under the reference scenario. However, the alternative scenarios proved that the supply of electricity in Nigeria could be reduced if proper energy efficiency policy and strategies are in place in the demand side.[34]

---

[33]In order words, people will purchase more private cars, commercial vehicle operators will increase their vehicle fleet since biodiesel becomes more available, and income increases at the same time.

[34]The discussion on electricity demand and supply will be given in the next subsection which deals with matching electricity demand with supply.

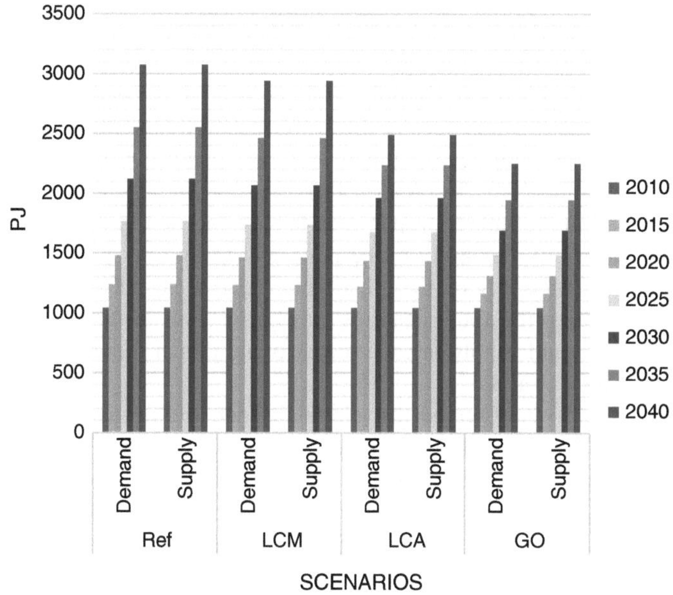

Fig. 5.32 Scenario's energy demand and supply

Natural gas supply for the reference scenario was 111.3 by 2040, and this was reduced as demand for the product declined in the LCM scenario (95.8 PJ). However, with increased use of natural gas in industry and power plants, the supply of natural gas to industry sector increased to 117.9 PJ, while power sector was supplied with 584.3 PJ by 2040. Figure 5.32 presents the various scenario energy demands to meet with supply.

The fuel supply shown in Fig. 5.32 covers all the fuels required in meeting the five sectors' energy demand. As observed in Fig. 5.32, whenever the demand in a scenario increases or reduces, so does the supply for the energy fuels. In a bid to understand this, a close observation of some fuels such as gasoline, kerosene, and diesel for example, had a supply of 470.4 PJ, 740.6 PJ, and 112 PJ respectively under the reference scenario in 2040. This however, was reduced in the alternative scenario where the demand for the fuel declined due to the impact of energy efficiency policies and fuel switching options in the sectors.

The supply of LPG in the Ref scenario was just 177.2 PJ due to low demand for the fuel in the household sectors. But in the alternative scenarios such as the LCA and GO, the supply increased due to the introduction of LPG fuel for vehicles and its increased use in the household sector. Biomass supply, which was 795.1 PJ in the Ref scenario had an increased supply in the LCM (880.1 PJ) scenario due to increasing demand in the transport sector. But this was reduced in the LCA (531.8 PJ) and GO (386.7 PJ) due to the introduction and increased share of LPG and CNG for the transport sector, and also other sectors such as household, industry, and commercial/service sectors.

5.2.2 Matching Electricity Supply with Demand

Matching Nigeria's electricity supply with demand has been a "tough-nut to crack" which is due to challenges in the power generation sector. This challenges include among others, improper planning of power plants construction and capacity expansion (Sambo 2008; Iwayemi 2008). However in matching electricity supply with demand in the future, proper knowledge is needed in order to understand the growing electricity demand on one hand and the proper generation capacity needed to supply electricity on the other hand (Ajao et al. 2009). To achieve this, a comparison among the four scenarios was carried out in order to observe the electricity demand and supply in each scenario and this is shown in Fig. 5.33 and presented in Table E.1 and E2.

From Fig. 5.33, it is observed that electricity demand in the Ref scenario will increase from 32,300 MWh in the base year 2010, to 161,800 MW by 2040. This demand will be matched with the supply of 179,800 MWh from electricity generation by 2040. Although the supply of electricity far exceeds the demand under the Ref scenario, it should be noted that the power plants proposed by the government will include more inefficient plants such as SCGT, gasoline, diesel, coal steam, and gas power plants that are also known for their high emission rate. On the demand side, policies were not in place under this scenario, while fuel switching was absent in all sectors in order to ensure maximum utilization of energy forms.

Moving to the LCM scenario, electricity demand (161,500 MWh) was not very much low due to the near absence of energy efficiency policies in the five sectors.

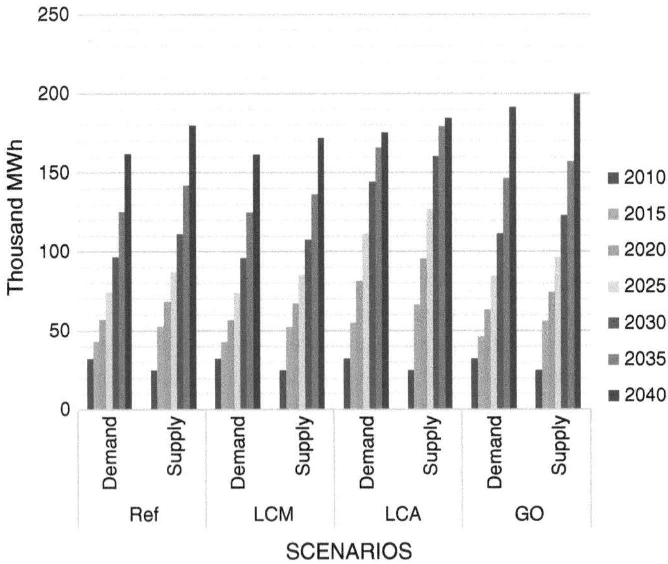

Fig. 5.33 Scenario's electricity demand and supply

However, electricity supply was still higher than demand but lower than the Ref scenario. More increase in electricity demand as compared to the LCM and Ref scenarios was observed in the LCA scenario with an increase of 175,100 MWh in electricity demand. This was attributed to not only the increase in income of households,[35] but the increased reliance of electricity as a form of energy preferred in the industry and commercial sector. However, the provision of solar electricity in households which is present even in the Ref scenario base year had its capacity increased and intends to reduce the dependency on grid supply in the demand sectors.[36]

The GO scenario which is focused on the increased use of renewable for electricity generation at the supply and to some extent, the demand side, had an increased demand for electricity to 191,400 MWh by 2040. This will be matched with the supply of electricity mostly from renewables and low carbon sources to produce up to 199,400 MWh by the end of 2040. As previously explained in Chap. 4 and the Results section in this chapter, off-grid electricity supply will come mainly from renewable energy sources with the exception of SCGT power plant to support electricity generation from renewables.

5.2.3 Primary Resource Requirements

In general, the primary resource requirement is expected to increase as each scenario changes its form of energy and electricity consumption. This is evidently shown in Fig. 5.34 and presented in Table D.1 which shows the value of primary resource requirement for each scenario from 2010 to 2040.

It can be observed that the LCA scenario presented a lower primary resource requirement when compared with other scenarios. This is due to the reliance on both conventional and renewable energy resources at an amount that is efficiently utilized for both power generation and energy consumption. This ensured the reduction of primary resource requirement of 3,507 PJ in the LCA scenario. The GO scenario also presented a low primary resource requirement of 3,507.5 PJ by 2040 which is better in terms of primary resource conservation as compared to the LCM and Ref scenario.

[35]This comes with household consumers purchasing more electrical appliances as reported in Dzioubinski and Chipman (1999), Robertson et al. (2011), and Abd'razack et al. (2013). More recently, Euromonitor International (2014) consumer lifestyle report on Nigeria population identified the growing relationship between increasing income and the behavior of consumers to acquire more goods to satisfy their needs. This includes household electronics and appliances which they intend to use in making their daily lives more comfortable.

[36]This in some point includes residential buildings in both the urban and rural areas, as well as commercial/service buildings such as banks, hospitals, government and agency buildings.

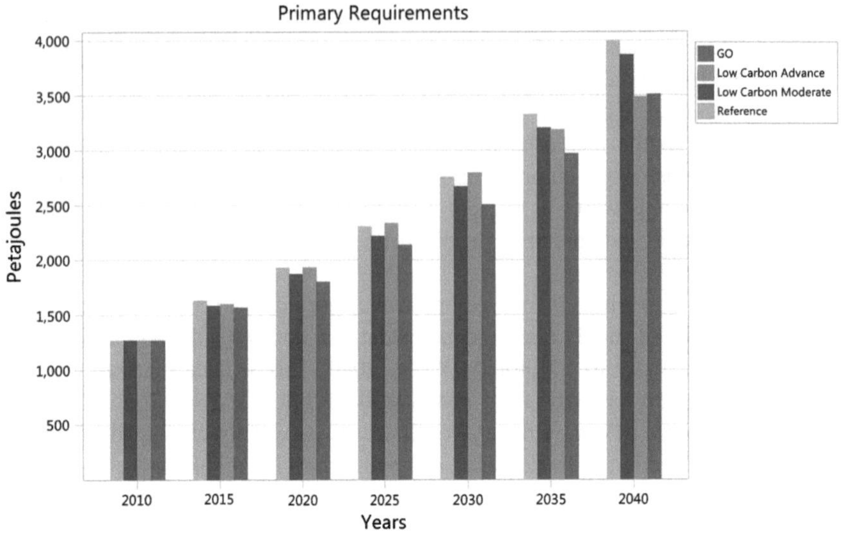

Fig. 5.34 Primary resource requirements

5.2.4 GHGs Reduction Potential

The GHGs recorded in the four scenarios considered revealed various levels of reduction in GHGs due to the various energy policies applied at different stages of the development of the Nigeria LEAP model. The summary of the GHGs from energy consumption and electricity generation is presented in Fig. 5.35. Figure 5.35 shows that in the Ref scenario, GHG emission will increase from 50.2 MMTCDE in 2010, to 201.2 MMTCDE by 2040. The next scenario which is the LCM scenario presented a slightly lower emission of 185.4 MMTCDE in 2040. It can be recalled that the energy policies and strategies implemented in the LCM scenario were moderate and not aggressive as the LCA and GO scenario.

As observed from Fig. 5.35, the LCA scenario developed initially had increased GHGs levels from 2020 to 2030 (10 years) after which the impact of energy policies began to reduce GHG emissions. It is, however, important to note that this reduction is not only from the GHGs from energy use alone, but also electricity generation. The GO scenario displayed a very low emission rate from 2015 to the target year, 2040. This was achieved due to the more aggressive policies in reducing energy demand on one hand, and improving the share of renewables on the other hand. This led to the reduction of GHGs to 124.4 PJ by 2040 from its 2010 levels. In terms of GHG emission savings, 15.8 MMTCDE was saved in the LCM scenario, 49.1 MMTCDE in the LCA and 76.7 MMTCDE in the GO scenario.

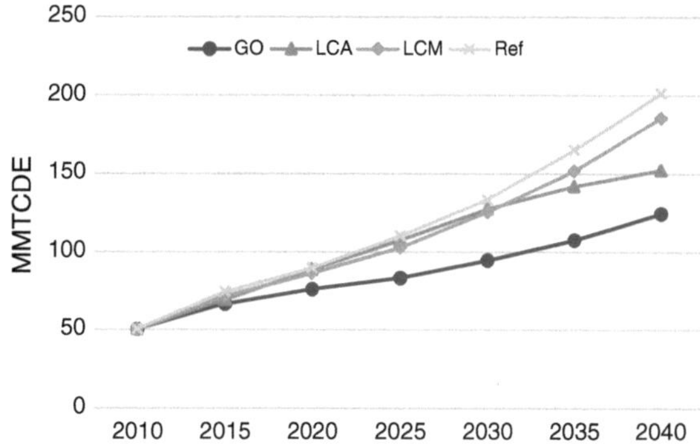

Fig. 5.35 GHG emissions summary for all scenarios

5.2.5 Cost-Benefit Analysis

A cost-benefit analysis or benefit-cost analysis can be described as a systematic approach to estimate the strengths and weaknesses of alternatives that satisfy transactions, activities, or functional requirements for a business. It is an analytical tool or technique used in the determination of options which provide the best possible approach in the adoption or practice in terms of benefits in labor, time, and cost savings (David et al. 2013). It can further be described as a systematic process observed in calculating and comparing the benefits and costs of a project, decision, or government policy. The aims of cost-benefit analysis include: to ascertain if an investment or decision is feasible, and to establish a basis for projects comparison (CEEU 2013; Policy 2014). This includes comparing the total expected cost of each option against the total expected benefits, in order to observe which benefits outweigh the costs and by how much (Spellman 2015).

The definitions provided above were intended to establish a general understanding of the term "cost-benefits analysis". The result of the cost-benefit analysis for the scenarios analyzed in this study is presented in Table 5.1. The results show the extra cost (if it is a positive cost) of implementing a scenario from the reference scenario; in other words, the differences in cost that remain when the reference scenario is subtracted from the alternative scenario(s).

In some situations, a benefit (if it is a negative cost) when the reference scenario is subtracted from the alternative scenario(s). The cost observed when a reference scenario is subtracted from the alternative scenarios is usually high or a positive cost. The net present value (NPV) of the alternative scenario as shown in Table 5.1 is relative to the reference scenario. The NPV is the sum of all discounted costs[37] and benefits in one scenario minus another[38] (i.e. LCM—Ref scenario or LCA—Ref scenario, etc.).

[37]This study assumed the discounted cost to be 5 % in 2010.

[38]This is summed-up across all the years of the study and in this study, it was summed from 2010 to 2040.

Table 5.1 Results of the cost-benefit analysis

Cumulative costs and benefits: 2010–2040. Relative to scenario: reference Discounted at 5.0 % to year 2010. Units: Billion 2010 U.S. Dollar	Low carbon moderate	Low carbon advance	Green optimistic
Demand	36.62	31.32	30.61
Households	0.42	0.56	0.57
Industry	–	–1.61	–2.35
Commercial_Service	–2.73	–5.74	–9.56
Agriculture	–	–	–
Transport	38.93	38.11	41.95
Transformation	15.65	36.12	81.62
Transmission and distribution	–	–	–
Electricity generation	15.65	36.12	81.62
Charcoal production	–	–	–
Oil refining	–	–	–
Coal mining	–	–	–
Resources	1,640.19	23,790.23	41,269.17
Production	–209.43	44.14	–303.14
Imports	1,849.62	23,746.09	41,572.31
Exports	–	–	–
Unmet requirements	–	–	–
Environmental externalities	–4.02	–24.05	–34.48
Non energy sector costs	–	–	–
Net present value	1,688.44	23,833.61	41,346.92
GHG savings (MMTCDE)	221.73	332.14	952.00
Cost of avoiding GHGs (U.S. Dollar/TCDE)	7,615.00	71,757.93	43,431.71

*Ton per carbon dioxide equivalent

From Table 5.1, it can be observed that under the demand variable (or demand side), the cost for the implementation of the LCM scenario is 36.62 million US dollar, while for LCA and GO scenario it was 31.32 and 30.61 million US dollar respectively. In implementing the polices in the LCM scenario, the benefit of 2.73 million US dollars will be saved by the Nigerian government, while 38.93 million US dollars would have been spent in achieving the policies in the transport sector and 42 million US dollars on the household sector. In the LCA scenario, more cost would be spent for household's energy efficiency policies to the tune of 560 thousand US dollars, while 38.11 million US dollars will be spent in the transport sector.

These will include the introduction and increase of LPG vehicles, energy efficiency, and fuel switching. The industry and commercial/service sectors in the LCA and GO scenario gained the benefits of 1.61 and 5.74 million US dollars in the respective sectors. The GO scenario showed a lower investment cost for the demand-side energy policy. Although the costs required to enable the policies in the GO scenario's household and transport sectors were high (i.e. 570 thousand US dollars and 41.95 million US dollars respectively), the benefits were observed in the industry and commercial/service sector (i.e. 2.35 and 9.56 million US dollars respectively).

The transformation sector showed various levels of costs which varied across scenarios. The LCM scenario had a lower amount of cost in terms of implementing the strategies,[39] while LCA had a higher cost of 36.12 million US dollars. The highest cost among the alternative scenarios was observed in the GO scenario, which is due to the high-capacity installation of the power plants which are mainly renewables. On resources, offset cost or benefits were observed for resource production in the LCM and GO scenario, while the LCA scenario costs about 23.8 million US dollars versus the Ref scenario. However, higher cost of import of energy resources was observed in the GO scenario as compared to the LCA and LCM scenarios. This high cost in the GO scenario was due to the increased importation of CNG, LPG, and nuclear fuels for the large capacity installed.

Environmental externalities[40] presented a benefit in all the alternative scenarios with the GO scenario having the highest benefit of 34.48 million US dollars, followed by the LCA scenario with a benefit of 24.05 million US dollars. The overall NPV for the alternative scenarios are 1.69 billion US dollars for the LCM, 23.8 billion US dollars for the LCA, and 41.4 billion US dollars for the GO scenario. This overall NPV is the cost of how much more the alternative scenario[41] costs versus the Ref scenario. The amount of GHGs saved in the LCM scenario was lower (221.73 MMTCDE) than the LCA (332.14 MMTCDE) and GO scenario (952 MMTCDE). Lastly, the cost of avoided[42] GHG emissions was higher in the LCA (71.8 US dollar/TCDE) scenario, than in the GO (43.4 US dollar/TCDE) scenario.

5.2.6 Least-Cost Electricity Generation

The power sector which has been built and operated under a "supply-follows-demand" philosophy has always been able to fulfil its obligation by providing adequate and secured supplies of electricity at the lowest practicable cost

[39]These strategies include the building and capacity expansion of power plants that are not in the Ref scenario. For some power plants in the Ref scenario that are in the alternative scenarios, the differences are taken which is the cost of the expanded capacity of the alternative scenario which is the difference between the two scenarios. Please note that the strategies also include the cost of improvement in T&D.

[40]Environmental externalities refer to the economic concept of uncompensated environmental effects of production and consumption that affect consumer utility and enterprise cost outside the market mechanism. As a consequence of negative externalities, private costs of production tend to be lower than its "social" cost. It is the aim of the "polluter/user-pays" principle to prompt households and enterprises to internalize externalities in their plans and budgets (United Nations 1997).

[41]Please note that it is one alternative scenario versus the Ref scenario (i.e. LCM vs. Ref, LCA vs. Ref etc.).

[42]This is given by dividing the NPV by the TCDE avoided.

(Santisirisomboon et al. 2003). The objective of the power generation plan[43] is to seek the most economical generation expansion scheme to achieve a certain reliability level to meet the forecasted demand increase in a certain period of time (Booma et al. 2006). But this has resulted in a strenuous operating policy/strategy that entails higher plant margins and extensive environmental impact in the energy system which invariably contribute to the increasing GHGs levels.

According to the U.S. Environmental Protection Agency (EPA), the power sector is the highest contributor to the global GHGs emission with about 26 % of the total GHGs by source (EPA 2015). Since this is the case, then the power sector has the highest potential in GHG mitigation. A lot of options exist in reducing GHG emission from the power sector and they include the transition from high to less or non-GHG intensive sources of electricity[44] generation (Sims et al. 2007).

This subsection however aims to evaluate the most economical power generation of 200,000 MWh by 2040 from the electricity power plant installed in 2010. It is important to note that the basis for the electricity generation limit of 200,000 MWh was due to the maximum electricity generation recorded in the scenario's results in the last subsection which was the GO scenario (i.e. 199,400 MWh). This subsection intends to set the limit to 200,000 MWh instead and then carry out a least-cost optimization with the energy technologies proposed in this study. It is however, important to note that the technologies do not include off-grid gas power plants, diesel and gasoline generators. Therefore, the power plants used in the least-cost optimization were biomass, CCGT, coal CFB, coal steam, coal supercritical, geothermal, large hydropower, nuclear, off-shore wind, on-shore wind, SCGT, small hydropower, solar PV, and solar thermal.

Each power plant was installed and increased to generate 200,000 MWh by 2040 from 2010, and this was done alone (i.e. not an electricity mix, but only that type of power plant was expanded). The methodology used for the least-cost electricity generation was presented in Sect. 4.1.2.6 in Chap. 4, while the data used are from Tables B.25 and B.26. The results are shown in Figs. 5.36, 5.37, 5.38, 5.39, 5.40 and 5.41 and presented in Tables I.1–I.6. In order to enhance the understanding of the results, the Figures are arranged in a way that each deals with a set of technology such as fossil fuel, low-carbon, renewable energy technologies, and specific GHG emissions.

5.2.6.1 Fossil Fuel Power Plants

The results for the least-cost electricity generation for fossil fuel power plants are shown in Fig. 5.36. It can be observed from the Figure that coal steam power

[43]In some circumstances, the plan translates to the least-cost electricity generation which seeks to minimize the total generation cost which is composed of generation capital investment costs, operation and maintenance (O&M) costs, outage cost, transmission losses costs and transmission enhancement costs (Sardou et al. 2013).

[44]Other measures include the application of Carbon Capture and Storage (CCS), installation of energy-efficient power plants with reduced emission rates (e.g. CCGT).

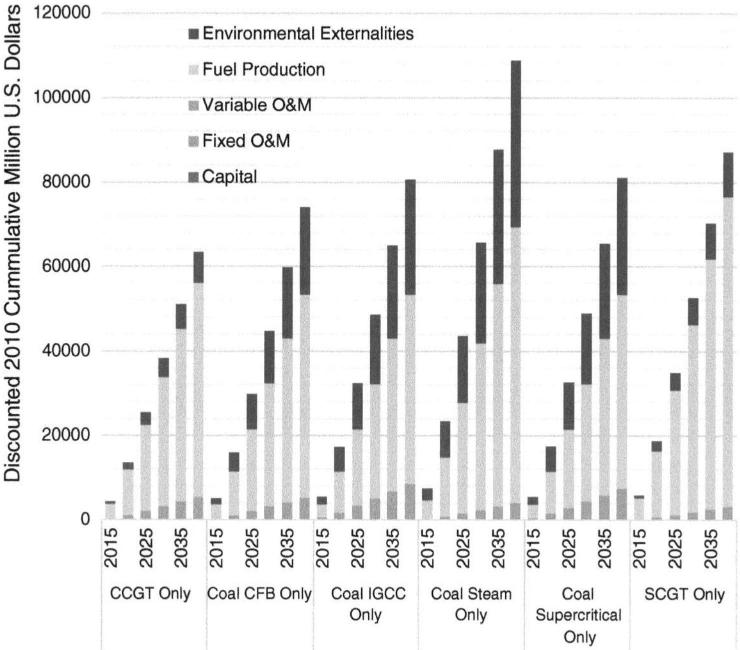

Fig. 5.36 Least-cost electricity generation for fossil fuel power plants

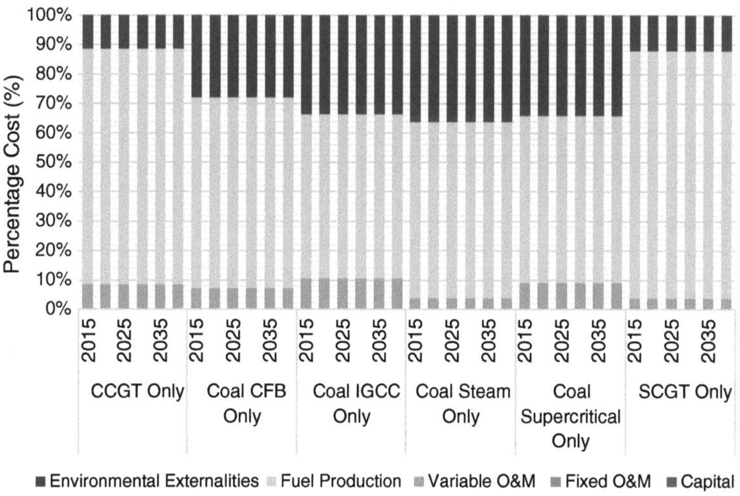

Fig. 5.37 Percentage share of costs for fossil fuel power plants

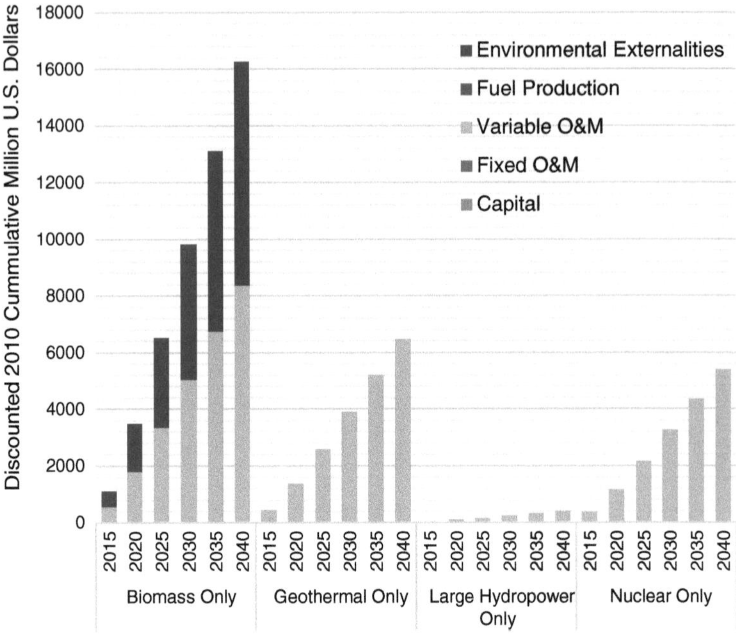

Fig. 5.38 Least-cost electricity generation for low-carbon power plants

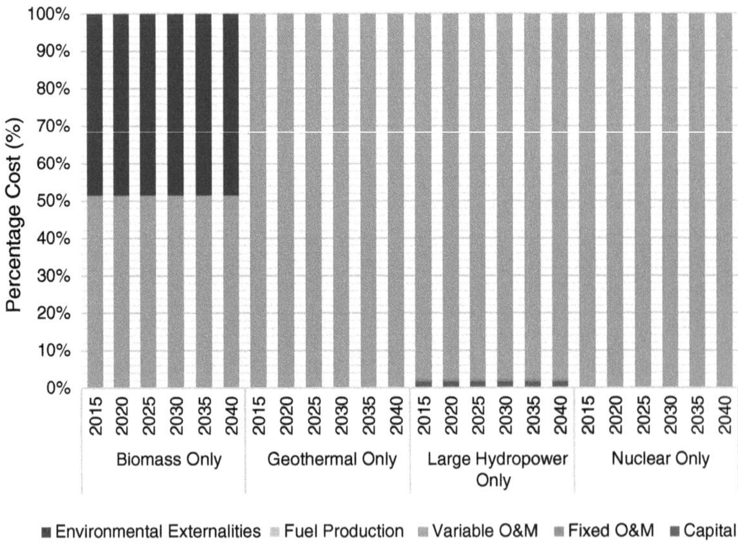

Fig. 5.39 Percentage share of costs for low-carbon power plants

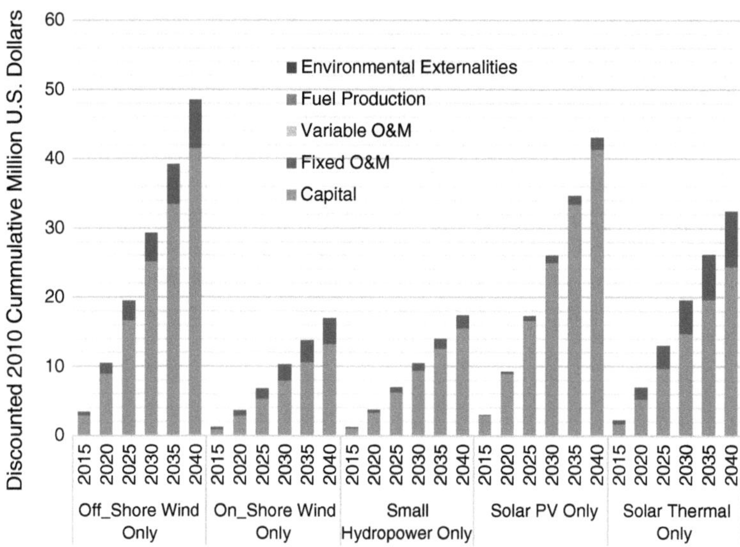

Fig. 5.40 Least-cost electricity generation for renewable energy power plants

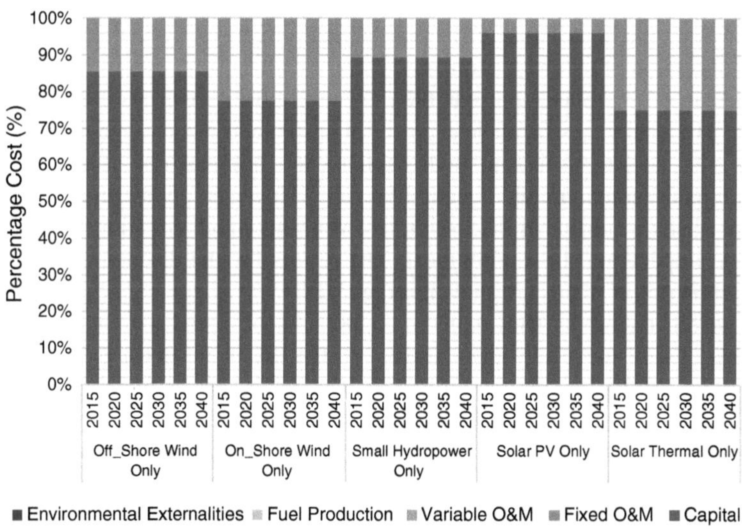

Fig. 5.41 Percentage share of costs for renewable energy power plants

plants had the highest cost for electricity generation in meeting the 200,000 MWh by 2040. Among the contribution to the high discounted cost recorded in the coal steam power plant was the cost of fuel production (i.e. cost of purchasing fuels for the power plants, in this case it was coal) which amounted to 65.3 billion US dollars.

The next pertinent issue is environmental externality cost which amounted to about 39 billion US dollars, while the total discounted 2010 cumulative cost by 2040 was 108.8 billion US dollars in 2040. However, among the other three coal power plants (i.e. CFB, IGCC and supercritical), CFB presented the least cost in terms of electricity generation of about 74 million US dollars by 2040.

The least-cost electricity generation from fossil fuel power plant came from CCGT plant which had a higher cost in fuel production or cost of purchasing natural gas to be 50.8 billion US dollars. The SCGT was higher than CCGT plants and even the coal power plants by 2040. Also in order to enhance the interpretation of the result, the share of cost (in percentage) in each power plant is shown in Fig. 5.37.

5.2.6.2 Low-Carbon Power Plants

These energy technologies are known to be power plants that emit an insignificant amount of GHGs or sometimes emit no GHGs at all. Large hydropower plants in real sense do not emit GHGs and are classified as renewable but have negative impacts on the environment. This sometimes leads to indirect emission of methane and carbon dioxide as in the case where decomposing plants and trees around the dams release methane and carbon dioxide which increase pollution. Other problems include the disturbance of natural habitat, effects on agriculture, breaking of dams, distortion of fish migration, etc.[45]

Nuclear power plants present their own disadvantages, despite not emitting GHGs. The challenges of nuclear power plant include nuclear accidents, radioactive waste, nuclear radiation, high capital cost, national risk, fuel availability, major impact on human and aquatic life, etc.[46] However, geothermal and biomass have not been a subject of discussion that warrants much attention based on their negative impacts on the environment. The results for the discounted costs in the low-carbon power plants are shown in Fig. 5.38.

From the results as shown in Fig. 5.38, it is observed that hydropower had the least-cost of electricity generation with the discounted cost of 400.9 million US dollars by 2040. This was followed by nuclear power plants with a higher variable O&M cost which increased its discounted cost to 5.4 billion US dollars. Comparable to nuclear power plant is geothermal with an increased discounted cost of about 6.5 billion US dollars. Nuclear power plant had a higher capital and O&M costs than geothermal power plant. Biomass power plants had a higher discounted cost of generating the 200,000 MWh of electricity by 2040. This was due to the higher variable O&M cost (8.34 billion US dollars) and externalities costs of about 7.9 billion US dollars. This is shown in percentage shares in Fig. 5.39.

---

[45]www.conserve-energy-future.com/Disadvantages_HydroPower.php.

[46]www.conserve-energy-future.com/Disadvantages_NuclearEnergy.php.

5.2.6.3 Renewable Energy Power Plants

Renewable energy technology presents the society with zero-emission power generation without fuel requirement, hence no environmental externalities. Among the renewable power plants proposed here includes off-shore and onshore wind turbine power systems, small hydropower, solar PV and thermal system, and large hydropower (as discussed in the last subsection) as these presented a lower discounted cost when compared to the fossil fuel and low-carbon power plants. For renewable energy technologies, on-shore wind power plant was observed to be the least-cost for electricity generation with a discounted cost of 17 million US dollars by 2040. The discounted cost for the renewable energy power plants is shown in Fig. 5.40.

This is closely followed by small hydropower plants with a discounted cost of 17.4 million US dollars, which differ from on-shore wind turbine power plants with 400 thousand US dollars. However, small hydropower had a smaller fixed O&M discounted cost of 1.8 million US dollars than on-shore wind turbine plants which was at 2 million US dollars. Off-shore wind turbine power plants were observed to have the highest discounted cost among the renewables proposed with a discounted cost of about 48.6 million US dollars by 2040. The percentage share of cost for renewables goes mainly to capital and fixed O&M cost as presented in Fig. 5.41. More capital cost is spent on solar PV system than other renewables in terms of share comparison, while solar thermal power plant has a lower capital cost.

5.2.6.4 Power Plant Specific GHG Emissions

The least-cost electricity results previously discussed explored the power plant options in order to identify the best option or power plant technology to achieve electricity generation of 200,000 MWh by 2040. The results showed that the least power plant was on-shore wind turbine power system which falls under the category of renewable energy technology, while CCGT power plants were observed to be the least-cost for fossil fuel power plants. This section briefly identifies the GHGs from each fossil fuel power plants as it increased from 2020 to 2040, and the result is presented in Fig. 5.42 and in Tables I.7–I.8.

Results of this study show that biomass is considered as a low-carbon energy technology emitting the least amount of GHGs across the years under observation. However, among the fossil fuel power plants, least GHGs were observed in CCGT power plants with an emission of 62.4 MMTCDE by 2040. It was also observed that carbon dioxide had the largest share in GHGs from the fossil fuel power plants. Coal power plants had higher emission rate as compared to gas power plants. Between the coal power plants, coal steam power plant was observed to emit high proportion of GHGs in order to generate the required 200,000 MWh by 2040. In conclusion, these results (including the least-cost electricity generation) intend to provide some policy guidelines in power plants investment in a way that it benefits the society in the short and long run. Investment in power plants should critically consider the long-term impact of each technology and not the low cost of investment on a short-term basis.

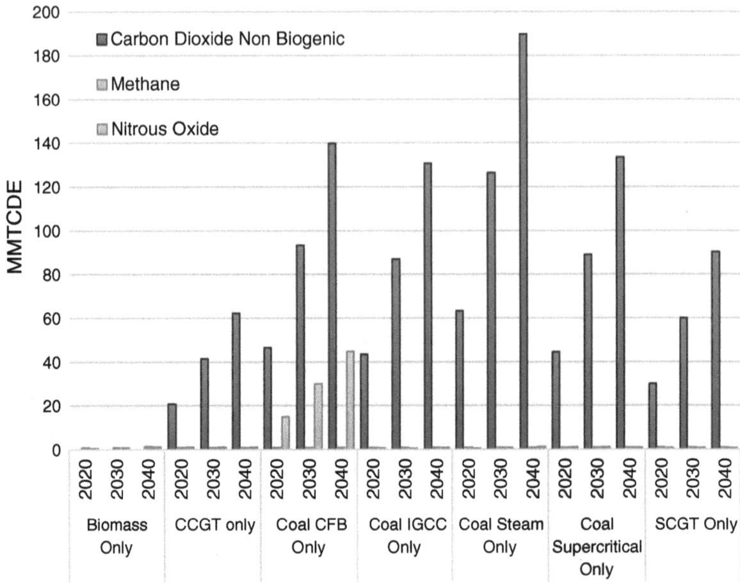

Fig. 5.42 GHGs in power plants

5.3 Sustainable Strategies

Attaining a low-carbon society requires the formulation and implementation of strategic policies that are sustainable in the long term.[47] Usually, low carbon development is followed by the move towards green growth which results in a green society (Wang et al. 2010). In a developing country such as Nigeria, green growth can be a kind of economic strategy to ensure sustainability in the presence of resource constraint and climate change (Emodi 2015). This section however explores the strategies that can ensure low carbon development in Nigeria with a view of attaining green growth.[48] The strategies are based on the results obtained in Sect. 5.1–5.2, and combined with relevant country cases.

[47]http://www.greengrowthknowledge.org/sites/default/files/kggp_knowledge%20note%20-series_01.pdf.

[48]According to the World Bank, green growth is the growth that is efficient in its use of natural resources, clean in that it minimizes pollution and environmental impacts, and resilient in that it accounts for natural hazards and the role of environmental management and natural capital in preventing physical disasters. The Organization for Economic Co-operation and Development (OECD) defines green growth as fostering economic growth and development, while ensuring that natural assets continue to provide the resources and environmental services on which our well-being relies.

5.3.1 Thinking and Adopting the Green Growth Ideology

Over recent years, the concept of "green growth" has been making waves in the international policy scene (Jacobs 2012). The green growth approach adopted by the Fifth Ministerial Conference on Environmental and Development (MCED)[49] sought to harmonize economic growth with environmental sustainability, while improving the eco-efficiency of economic growth and enhancing synergies between environment and economy. As with green economy, green growth attracted significant attention as a way out of today's economic doldrums in the aftermath of the 2008 financial crisis.[50] In 2008, the Korean Government adopted "low carbon green growth" as the country's new development vision in response to the global financial crisis (SDKP 2008).

The advancement to a low-carbon economy and then green growth should start in the manner of thinking of not only the Nigerian government, but all and sundry in the country are to be involved in this development. It is not enough to say that Nigeria is the largest economy in Africa (World Bank Databank 2015a, b), while energy demand far exceeds supply. Nigeria needs to focus on shifting from quantity of economic growth to quality of economic growth. In adopting a green growth strategy, Nigeria will not only change the way its cities, industries and infrastructures are being built, but also open up opportunities for income generation. As observed in the GO scenario for the household sector, the provision of solar PV systems for household will enable electricity consumers to become electricity generators who can sell power back to the grid[51] ("Prosumers"). This can be done through feed-in-tariff[52] (FIT) for potential household owners, while electricity generators can benefit from renewable portfolio standard[53] (RPS) as observed in the GO scenario for electricity generators.

Ignoring the idea of low carbon-green growth leads to the increased cost of climate change and other natural disaster on economic growth. This has direct impact on a country's GDP as natural disasters such as floods due to heavy rains and rising sea levels, heat waves, desert encroachment, and deforestation alter economic

[49]This conference was held in March 2005 in Seoul, Republic of Korea.

[50]It is important to note that the concept of green growth has its origins in the Asia and Pacific Region.

[51]Grid-extension is often the least-cost option for electrifying yet-to-be-electrified communities and off-grid only becomes viable if the communities are too remote and the demand level is relatively very low. If off-grid solar PV system is then used to electrify such communities, they may not be able to sell power back to the grid, but a proper mini-grid system can ensure power is sold within the mini-grid in the communities.

[52]FIT is a policy mechanism designed to accelerate investment in renewable energy technologies through the offering of long-term contracts to renewable energy producers, typically based on the cost of generation of each technology (Couture et al. 2010).

[53]RPS is a regulation that requires the increased production of energy from renewable energy sources, such as wind, solar, biomass, and geothermal (National Renewable Energy Laboratory [NREL] 2015).

activities.[54] Thinking green growth while implementing low-carbon strategies will not only solve the energy poverty situation in Nigeria, it will also create job opportunities which will reduce the unemployment rate in Nigeria which still stands at 7.5 % (World Bank Databank 2015a, b).

5.3.2 Energy Policy Reforms

The Nigerian Energy Policy which is also known as the National Energy Policy, was developed to ensure optimal, adequate, reliable and secure supply of energy and its efficient utilization, while ensuring a sustainable development (NEP 2013). It is vital for policy reforms in the energy sector in order to keep abreast of the latest developments going on in the field of energy technology. The introduction and deployment of the suggested low-carbon technologies used in the scenario analysis (Chap. 4 and Sects. 5.1 and 5.2) can only be actualized through energy policy reforms.

Energy policy reforms will provide a historic opportunity to identify and revitalize some ailing areas in the energy sector and bolster the overall economy. In countries such as Mexico (Romero and Castro 2008), the United States of America (Joskow 2000), some European countries (Glachant 2009), South Korea[55] and other Asian countries,[56] energy policy reforms have played important roles in low carbon-green growth. Although the Nigerian government has made some reforms in the energy sector (Usman and Abbasoglu 2014), the reforms take considerable time in implementation and the effects of most energy policies take more than 10 years to be felt. The solution to these challenges would be to reduce the lifetime of most policies and set up policy evaluation units, with the responsibility of evaluating the effectiveness of energy policies in the county.

5.3.3 Long-Term Energy Planning and Target

According to Winston Churchill during the Second World War, "He who fails to plan is planning to fail".[57] A country that has no proper plans on energy for the future is bound to experience energy crisis.[58] Long-term energy planning can not only help mitigate shortage in energy supply, but also reduce GHGs in the environment.

[54]http://www.theguardian.com/environment/blog/2014/jul/14/8-charts-climate-change-world-more-dangerous.

[55]https://www.iea.org/publications/freepublications/publication/Korea2012_free.pdf.

[56]http://www.adb.org/sectors/energy/issues/sector-governance-reform.

[57]http://www.in.gov/dnr/water/files/wa-IEAP.pdf.

[58]http://www.thehindubusinessline.com/opinion/we-lack-a-clear-view-on-energy-planning/article6875164.ece.

Lack of proper planning can lead to pollution problems experienced in places such as Baoding in China[59] and some cities in India.[60] From the GHG emissions in the alternative scenarios (i.e. LCM, LCA and GO), various reduction potentials can be achieved if the government strategies are in place. For example, Fig. 5.35 shows reduced emission in the GO scenario as compared to the Ref scenario. This leads to the GHG savings as shown in result of the cost-benefit analysis.

Long-term energy planning and target creates a vision and strategy which has the capacity to promote technological innovation and development in low-carbon technologies (Emodi et al. 2015c). The research, development and deployment (RD&D) of low carbon technologies require considerable time for both research and development (R&D), commercialization and funding. Therefore, the Nigerian government should not only plan and set targets for energy development, but also set up important sustainable infrastructure development. This will stimulate investment and improve on innovative ideas between the government and private investors. An example is cited in South Korea where the government implemented the Five Year Green Growth Plan to stimulate the needed investment and innovations in low-carbon technologies (Mathews 2012; Moon 2010; Kim et al. 2011).

5.3.4 Energy Regulations and Standards

In moving towards a low carbon-green growth society, the Nigerian government should improve on its energy regulations and standards. This includes technical regulations and voluntary standards as being handled by the Standard Organization of Nigeria (SON).[61] However, renewable energy standards have not been adequately developed in Nigeria and this is vital in ensuring the acceptance of renewables in Nigeria (Emodi et al. 2014). The improvement on energy regulations and standards will provide an enabling environment and incentives, increase renewable energy consumers and business owner's confidence in deployment, and encourage innovation and commercialization of low carbon-green technologies (Bina 2013; Gibbs and O'Neill 2014; Bailey and Caprotti 2014; Linnér and Selin 2013). The realization of the alternative scenarios (especially the GO) will require a stronger commitment by the Nigerian government in improving energy regulations and standards. This includes a range of standards such as emissions, energy efficiency, fuel efficiency, energy efficiency building in residential, commercial, and industrial buildings. The effects of these policies were observed in the scenario analysis (see Sect. 5.2).

---

[59]http://www.theguardian.com/cities/2015/may/22/baoding-china-most-polluted-city-air-pollution-beijing-hebei.

[60]http://gizmodo.com/indias-air-pollution-is-so-bad-its-causing-lung-damage-1707775668.

[61]http://services.gov.ng/son.

An example of standards that can be enforced as a Law is the National Appliance Efficiency Standards Law. This law would ensure that every electrical appliance, either imported or manufactured in the country, would meet the minimum efficiency standards for energy consumption. The standards would cover all major household appliances (home entertainment appliances, refrigerators, freezers, washing and drying machines, electric cookers, and air conditioners), lighting products (lamps and fluorescent lighting ballasts), and other appliances used in the industrial, commercial, and service sectors of the economy (Geller 1997).

5.3.5 *Environmental Tax Reforms*

Environmental tax reform is defined as the reform of the national tax system where there is a shift of the burden of taxes, for example from labor to environmentally damaging activities such as unsustainable resource use or pollution (EPA 2015). This actually means shifting the tax base to resource consumption while maintaining revenue equality which will ensure the non-increase of the overall burden of tax. This has great potential in boosting economic growth and providing employment by reducing labor cost. This has great potential in boosting economic growth and providing employment by reducing labor cost. This is done by increasing the prices of energy resource use for electricity generation, transport, household, etc. and pollution by their accompanying tax. The tax revenues can be used to complement those taxes realized from the productive sectors of the economy (Ekins and Speck 2011).

This will ensure the effectiveness of policies proposed in the alternative scenarios such as fuel switching and fuel economy in the transport sector, conversion from conventional blast furnace to electric arc blast furnace in industries (See Sect. 5.1.4), and the diversification of electricity generation source such as renewable energy. Some countries have introduced environmental tax; these include India (Shukla and Dhar 2011), Iran (Reza et al. 2011), China (Liang et al. 2007), Canada (Harrison 2012), Indonesia (Yusuf 2008) and some European countries (Barker et al. 2007). These countries recorded some considerable amount of success in their environmental tax reforms, and this can be replicated in the Nigerian case if properly implemented.

5.3.6 *Urban Planning*

The increase in urbanization rate as assumed in Table 4.1 in all the scenarios calls for proper urban planning in cities and urban areas. This increase will also lead to increase in car ownership by the populace. Most cities and urban areas in Nigeria lack proper strategic planning (Innocent 2013). The proper planning of cities with a variety of mass transit options[62] and mixed use (as described in the GO scenario) can reduce energy demand and GHGs emission in the transport sector. The design of urban centers should emulate compact city[63] in order to reduce the need for mobility through transport. This is done by concentrating and mixing offices, commercial and residential areas, and ensuring walkable streets which are well connected are provided (Daramola and Ibem 2010).

In order to discourage private vehicle use, the Nigerian government should set up parking management authorities in urban areas (Jordan and Infante 2012). This approach was applied in Abu Dhabi which had problem of traffic congestion and increased pollution due to large number of private cars used. The government through the Abu Dhabi Urban Planning Council initiated the Abu Dhabi Urban Street Design Manual with guide design of creating walkable streets to encourage people to walk instead of driving (Abu Dhabi Urban Planning Council 2015).

5.3.7 *Efficient Building Design*

Reduction in energy consumption in residential, commercial and industrial buildings can effectively reduce the total demand for energy and reduce GHG emission in Nigeria. As discussed in Sects. 5.1.4.3 and 5.2.3, great potentials exist in energy saving in buildings if energy efficiency policies are rigorously followed. Already existing buildings in Nigeria will have to improve on energy efficiency, while new building designs should be based on low-carbon building standards.

According to studies by (Change 2007; Stern 2007; Davidson et al. 2007; Intergovernmental panel on Climate change 2007), buildings consume about 40 % of the global energy, 12 % of freshwater use and contribute 40 % of the total GHG emission in the world. The scenario analyzed showed that the household sector in the Ref scenario (see Fig. 5.15), energy demand will increase up to 3,075 PJ by 2040 if effective policy measures are not put in place. This is by far a high demand for energy as compared to the alternative scenarios. This shows the need for the development of efficient buildings so as to reduce demands for energy.

Some challenges that may arise in the development of efficient building designs include lack of public awareness, access to financial resources and the unavailability of new efficient technologies which may be costly to the users.[64] The introduction of energy labeling program in the LCA scenario also presents its potentials for not only efficient appliance use, but also to save utility bills by the consumers.

---

[62]This includes the increased access to mass transit use and the discouragement of private car use as was discussed in the LCA and GO scenarios for the transport sector.

[63]Measures include the promotion of vertical and cluster development, encouragement of cellular development, and setting urban growth boundaries to limit urban and suburban sprawl.

[64]These challenges are supported in Majid and Hussaini (2011) study who through a household opinion survey of Nigerian households were able to identify the challenges of energy efficiency in Nigerian buildings in relation to the building designs.

The Nigerian government can use regulatory measures such as building certificate and audit programs, building codes, and appliance standards for high efficiency to aid in improving energy efficiency in buildings. Other means that can be used by the government include fiscal incentives, which can come in forms of grants, subsidies, and tax exceptions which are awarded to residential, commercial, and industrial buildings employing energy efficiency measures. This will be useful in encouraging industries in acquiring electric arc blast furnace as applied in the GO scenario (see Sects. 4.3.5 and 5.1.4).

5.3.8 Efficiency of the Energy System

In order to advance in a low carbon-green economy, the efficiency of the Nigerian energy system needs to be improved. This will facilitate the production and consumption of low GHG emission and energy savings. As developed in the GO scenario, the shifting from fossil fuels to renewable energy options is vital in ensuring sustainability of power supply. In connection to improved urban planning and efficient building design, the demand-side management systems and energy efficiency will contribute to the realization of a low carbon-green economy in Nigeria.

Investing in low carbon and green technology is expensive as discussed in Chap. 5. The positive effect of this is the reduction in cost of investment overtime (Sect. 5.2.4). According to the International Energy Agency (2011), for every US$1 of investment in low carbon energy technology not invested before 2020, an additional cost of US$4.3 will be required to off-set the GHG emissions. This means that if the Nigerian government refuses to take actions now in investing in low carbon, energy efficiency and green technologies, the government will spend a lot out of its budget on climate disaster in the medium and long terms.

Other areas of the energy system that require efficiency improvement include T&D as discussed in the alternative scenarios. The diversifications to renewables should not only be at the supply side only, but also at the demand side, especially in the household sector (as observed in the GO scenario). In energy system, utilization of urban waste for energy generation should be incorporated as this will reduce pollution (urban waste) and contribute to electricity generation (through biomass power plants).

5.3.9 Efficiency of Passenger Transport System

The transportation sector is one of the main sources of energy consumption and environmental pollution in Nigeria. A more sustainable transport system could be achieved if there was proper coordination of urban and transport planning to reduce light vehicle and fuel use. Policy options for a sustainable transport system include the stimulation of public transportation through the introduction of an

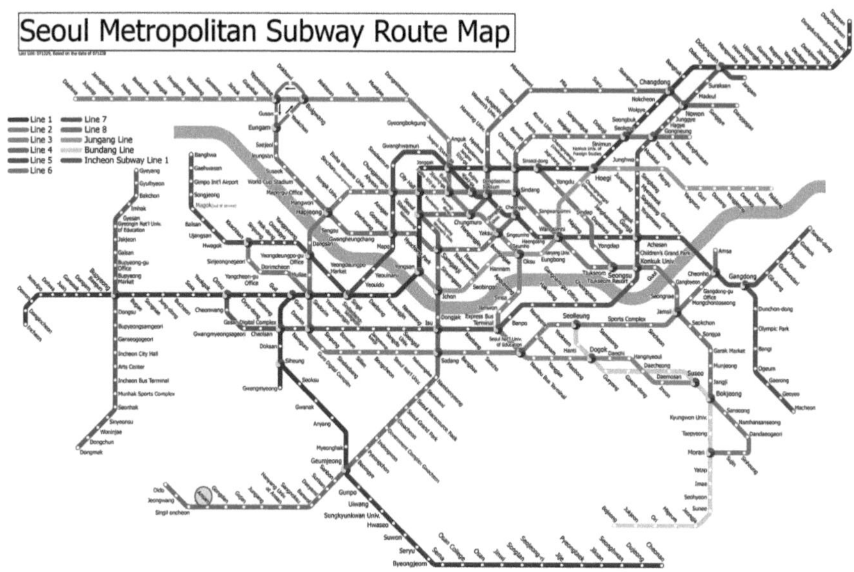

Fig. 5.43 Seoul metropolitan subway map. *Source* Visit Seoul (2015)

intermodal passenger and freight transport system that will better integrate and improve the lives of Nigerians. A specific example is the Seoul Metro System in Seoul,[65] South Korea, as shown in Fig. 5.43.

This subway system can reduce energy use in cars and buses for urban movement. For the Nigerian government to facilitate and implement this option, some strategies should be explored. These strategies include the promotion of subway use, expansion of the public transport infrastructure, use of urban tolls, increase in the cost of automobile parking spaces in congested urban areas,[66] restrictions on the use of automobiles to reduce congestion and air pollution in major cities, and improvements in telecommuting policies in Nigeria. A summary of the proposed strategies in attaining low-carbon development in Nigeria is shown in Fig. 5.44.

[65]The Seoul Metropolitan Subway is known as the world's longest multi-operator metro system by route length, and consists of 18 lines that serve the Seoul Metropolitan Area. The subway system is packed with advanced technology such as 4G LTE, WiFi, DMB, and WiBro services. Most of the trains are equipped with digital TV screens; all the trains are air conditioned, and include climate-controlled seats that operate automatically during the winter. All the subway lines use the T-money smart payment system using RFID and NFC technology for automatic payment by T-money smartphones or credit cards, and one can transfer to any of the lines within the subway system for free.

[66]These could be through a parking management policy as discussed in Johansson and Nakićenović (2012), and Jaccard et al. (2012).

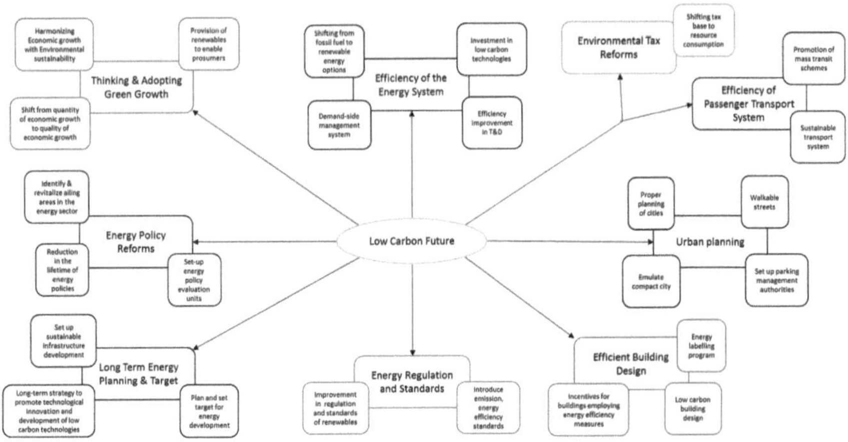

Fig. 5.44 Summary of low-carbon strategies for Nigeria

5.4 Summary and Policy Implication

5.4.1 Summary and Assessment

Based on the scenarios' assumptions of the socio-economic development and assumed parameters for the Nigerian LEAP model, the energy implication of the four possible pathways for Nigeria were analyzed using the LEAP model for the Ref, LCM, LCA, and GO scenarios between 2010 and 2040. The key findings of the analysis in this study are:

- Energy demand in the Ref, LCM, LCA and GO scenarios are expected to grow at annual rates of 3.58, 3.53, 2.95, and 2.61 % respectively. The energy demand under the GO scenario will be 2,249.2 PJ and this is the lowest among the four scenarios (see Fig. 5.45). The increase in economic activity, population, households, income, industry and transport needs are the major drivers in Nigeria's energy demand.
- The share of energy demand for the households under the Ref scenario is expected to increase from 40 % in 2010 to 42 % by 2040 (see Fig. 5.46). The energy demand in industry is expected to decrease to 26 % by 2040 from its share of 29 % in 2010. The share of energy demand in commercial/service will increase from 9 % in the base year, to about 16 % in by 2040. In the alternative scenarios by 2040, households' share of energy demand decreases to 39, 38 and 37 % in the LCM, LCA, and GO scenarios respectively. Also, the share of industry sector decreases across the alternative scenarios, while the transport and commercial/service sector had its share increased in LCM (22 %), LCA (23 %) and GO (23 %). The energy consumption of the residential, transport, and commercial/service sectors defined the vital role played by these sectors in shaping Nigeria's overall energy pathways.

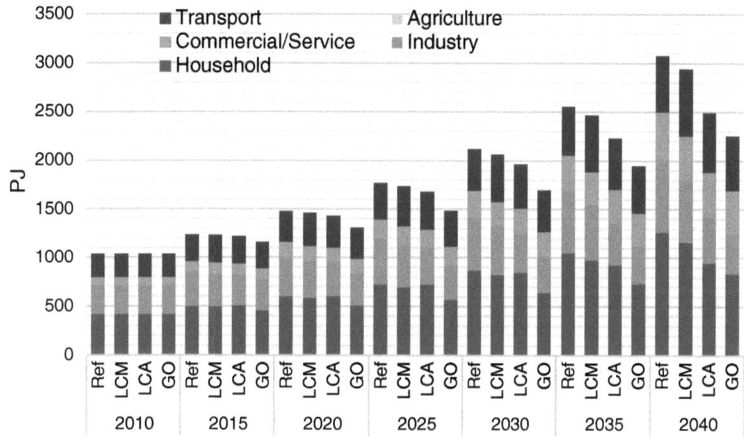

Fig. 5.45 Total energy demand comparison among scenarios' results

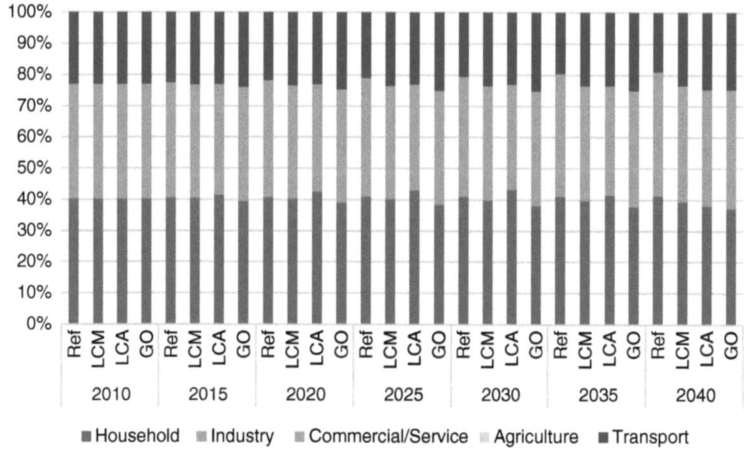

Fig. 5.46 Share of final energy demand by sector

- The total energy supply increases from 2010 to 2040, but decreases across the scenarios. In the Ref scenario, the share of electricity supply increases from 11 % to about 19 %, while biomass and kerosene had a share of 25 and 26 % respectively by 2040. Electricity supply increases in the alternative scenarios with the highest share observed in the GO scenario (see Fig. 5.47). Petroleum products (e.g. kerosene and gasoline) and biomass played a significant role in Nigeria's energy mix in all scenarios.
- The share of LPG increases from 5 % in 2010 to 13 % by 2040 under the LCA scenario, while CNG increases to 4 % by the end of 2040 in the GO scenario. Gasoline will have a reduced share in all alternative scenarios with the lowest

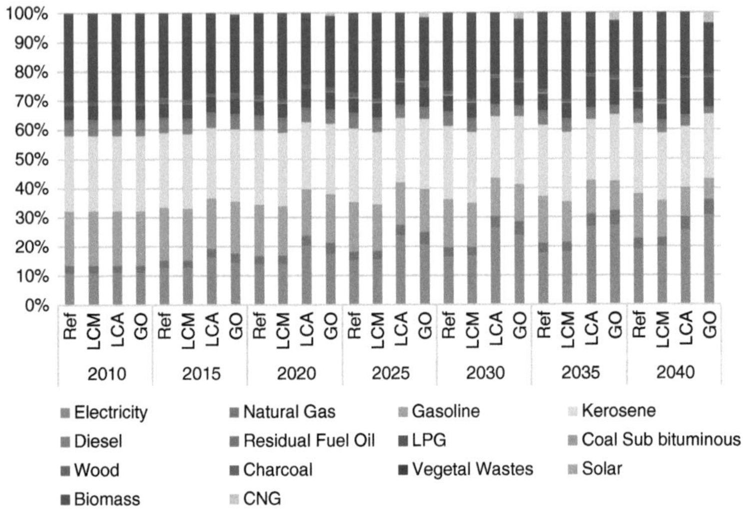

Fig. 5.47 Share of energy requirements by fuels (in demand sectors)

share of 7 % in the GO scenario. These fuels are expected to play an increasingly important role in the transport sector. The promotion of LPG, CNG and biofuel will provide a means of alternative fuel options in the transport sector.

- Electricity and natural gas T&D losses were reduced to 10 and 20 % in the Ref scenario, 6 and 13 % in the LCM, 5 and 10 % in the LCA, and 4 and 6 % in the GO scenario respectively. Improvement in electricity T&D would increase power supply to the demand by various sectors, while natural gas T&D improvement will ensure its sufficient supply to power plants and the industry sector.
- On electricity generation, the share of fossil fuel power plants will increase in the Ref and LCM scenarios as their electricity mix gets expanded. The LCA assumes an increased electricity supply from low-carbon technologies such as nuclear, CCGT and coal CFB, while balancing the mix with renewables. The GO scenario considers an increased share of about 70 % installed capacity from renewables (see Fig. 5.48).
- The structural changes in the electricity mix as shown in Fig. 5.48 mean that consumption of fossil fuel will reduce in the LCA and GO scenarios. This will occur as the energy usage structure gradually becomes more environmentally friendly in Nigeria, with a rapid growth in the use of low-carbon-green technologies.
- The overall GHGs emission in each scenario in this study is expected to increase from 2010 to 2040 (see Fig. 5.35). The Ref scenario had the highest GHG emission level of 201.2 MMTCDE by 2040, while a lower amount was observed in the LCM (185.4 MMTCDE), LCM (152 MMTCDE) and GO (124.4 MMTCDE). Although Nigeria is not under any obligation to reduce GHG emission since its emission level is very low, the global awareness of climate change is taken into consideration in the LCA scenario and more aggressively in the GO scenario.

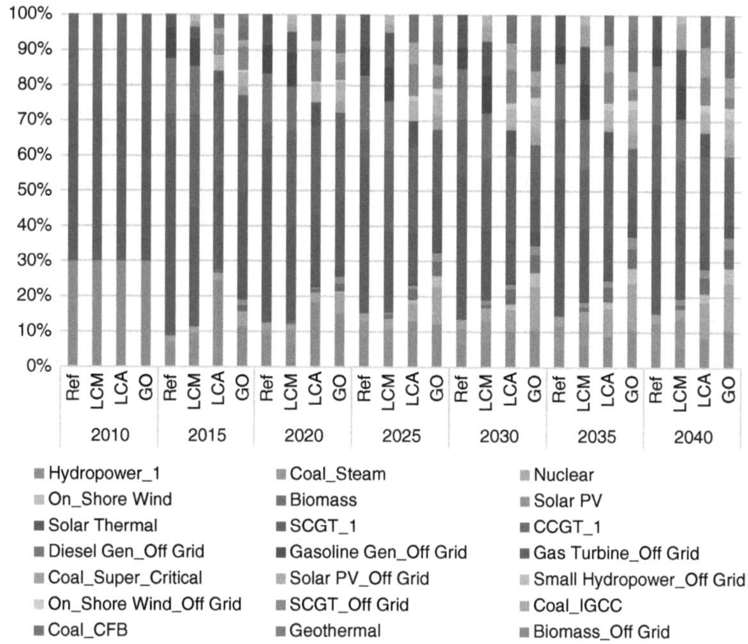

Fig. 5.48 Share of electricity generation mix

- Nigeria will continue to import petroleum products (petrol and kerosene) and gas products from overseas refineries in the long run due to increase in energy demand and low refinery capacity of domestic refineries. Thus, the government needs to build more refineries and identify more options in order to increase the production capacity of the Nigerian refineries.
- The cost-benefit analysis showed that among the three alternative scenarios as mitigation to the Ref scenario, the LCA and GO scenarios had a higher cost compared to the LCM scenario. The overall net present value (NPV) of the LCM costs 1.69 billion US dollars, LCA costs 23.8 billion US dollars, while the GO scenario costs 41.4 billion US dollars. Although these cost are high since investments are made in energy efficiency, fuel switching, etc., this is somewhat offset by energy savings on the demand-side, production, and environment externalities.
- On the least-cost electricity generation, on-shore wind power was observed to be the overall least-cost electricity generation option, while CCGT was the least cost for fossil fuel power plants. The optimization was carried out from 2010 to 2040, with an electricity generation target of 200,000 MWh.
- The power plants specific GHG emissions showed that coal steam had the highest GHGs emission by 2040, while the least GHGs came from CCGT power plants. Although biomass was included, its GHGs were insignificant; besides, the power plant was classified as a low-carbon energy technology.

In summary, the scenarios developed in this study present possible energy pathways and their impacts on the energy systems in terms of energy demand and supply mix, energy related GHG emissions, and cost-benefit analysis. Further, the study examined the least-cost electricity generation from the power plants used in modeling the electricity sector in the Nigerian LEAP model. These results are very useful for energy planning and the formulation of effective energy policies in Nigeria and other developing countries alike.

5.4.2 Policy Implication

Based on the results of this research, a number of policy implications can be drawn from the alternative scenarios and least-cost electricity generation in order to develop sustainable strategies and energy policies for the energy sector in Nigeria.

The first is improving energy efficiency in both the supply and demand sides through the application of modern energy technologies and practice. On the demand side, the implementation of energy efficiency practice such as replacement of incandescent bulbs for lighting with CFL and LED bulbs provides great energy savings potentials. The implementation of these policies should not only be focused on the residential sector, but also the industry and commercial/service sectors. This practice should also be extended to the provision of efficient cooking stoves since cooking and lighting are the most energy-intensive activities in the residential sector. Initiating an energy efficiency labeling program for household electronics and appliances such as air conditioning units and refrigerators will also make a substantial contribution to reduction in energy demand. On the supply side, efficiency improvement should be focused on the transmission and distribution network in both electricity and natural gas supply. Further, the oil refineries will need an improved energy efficiency practice and not just an increase in production capacity.

Secondly, the government should make provision for the development and deployment of alternative fuels in the demand and supply sides. This will ensure a decrease in dependence on a particular fuel source such as biomass and oil as seen in this study. This can be through the promotion of LPG for both residential consumption in cooking and lighting, and also in the transportation sector. Besides LPG as an alternative fuel, CNG and biomass options should be further explored to reduce oil dependency and limit GHG emissions. The government in conjunction with the private sector can work together in ensuring the deployment of alternative technologies that are efficient in production, such as electric-arc blast furnace for the industry sector. Also, cooperation will be needed in the adoption of CNG and LPG vehicles for transport fleet operators who may find it challenging in technology switching. This can be achieved through the implementation of a National Autogas Incentives and Policy to stimulate stakeholders in developing the sector.

Thirdly, the government should encourage the use of renewable energy technologies in both the supply and demand side through the provision of incentives such as investment tax credits, low-cost loans for power generation or feed-in-tariffs. This should be done considering the importance of reducing fossil fuel demand, GHG emission and ensuring sustainability of the energy system. Although the cost of implementing energy policies and strategies is very high as shown in the cost-benefit analysis, offset cost could be realized from energy savings, reduction in the depletion rate of natural resources and GHGs. However, the possibility of the application of carbon capture and storage (CCS) exists for the coal and gas power plants, but this may be too early as the Nigerian government may not be too concerned about the CCS technology. Thus, the selection of low carbon technologies proves to be the most possible pathway in ensuring low GHGs levels.

Finally, the results for the least-cost electricity generation showed that CCGT power plants are the best option for electricity generation for fossil fuel power plants, while on-shore wind and small hydropower were found to be the least-cost among the renewables. This serves as an implication for the government and policy makers in ensuring the long-term benefits of utilizing the maximum potentials of these power plants. However, other energy technologies such as biomass, large hydropower, and nuclear still present some low-cost potentials, and renewables still offer better alternatives in power generation. In all, policy implementation strategies should be focused on achieving a long-term impact on the energy system which will positively benefit both the society in particular and the environment in general.

References

Abd'razack NTA, Ludin ANM, Umaru ET (2013) Ecological footprint, lifestyle and consumption pattern in Nigeria. Am-Eurasian J Agric Environ Sci 13(4):425–432

Abu Dhabi Urban Street Design Manual. Abu Dhabi Urban Planning Council. http://www.upc.gov.ae/guidelines/urban-street-design-manual.aspx?lang=en-US. Accessed 10 Sept 2015

Ajao KR, Ajimotokan HA, Popoola OT, Akande HF (2009) Electric energy supply in Nigeria, decentralized energy approach. Cogeneration Distrib Gener J 24(4):34–50

Akinbami JF, Lawal A (2009). Opportunities and challenges to electrical energy conservation and CO_2 emissions reduction in Nigeria's building sector. In: In Fifth Urban Research Symposium, Cities and Climate Change: Responding to an Urgent Agenda, pp. 28–30

Bailey I, Caprotti F (2014) The green economy: functional domains and theoretical directions of enquiry. Environ Planning A 46(8):1797–1813

Barker T, Junankar S, Pollitt H, Summerton P (2007) Carbon leakage from unilateral environmental tax reforms in Europe, 1995–2005. Energy Policy 35(12):6281–6292

Bina O (2013) The green economy and sustainable development: an uneasy balance? Environ Planning C: Gov Policy 31(6):1023–1047

Booma J, Mahadevan K, Kannan S (2006) Minimum cost estimation of generation expansion planning incorporating wind power plant

British Broadcasting Corporation News (BBC) (2012) Brazil dam burst forces thousands from homes. http://www.bbc.com/news/world-latin-america-16426732. Accessed 24 Aug 2015

Central Expenditure Evaluation Unit (CEEU) (2013) Guide to Economic Appraisal: Carring out a cost benefit analysis. The Public Spending Code. www.publicspendingcode.per.gov.ie/wp-content/uploads/2012/08/D03-Guide-to-economic-appraisal-CBA-16-July.pdf. Accessed 29 Aug 2015

Change IC (2007) Mitigation of climate change. Summary for Policymakers, 10

Couture TD, Analytics E, Cory K, Kreycik C, Williams E (2010) A policymaker's guide to feed-in tariff policy design
Daigneau E (2013) Is Hydropower a Renewable Energy or Not? Governing: Infrastructure and Environment. www.governing.com/topics/transportation-infrastructure/gov-hydropower-renewable-energy.html
Daramola A, Ibem EO (2010) Urban environmental problems in Nigeria: implications for sustainable development. J Sustain Dev Afr 12(1):124–145
David R, Ngulube P, Dube A (2013) A cost-benefit analysis of document management strategies used at a financial institution in Zimbabwe: a case study: original research. South Afr J Inform Manage 15(2):1–10
Davidson O, Bosch P, Dave R, Meyer L (2007) Mitigation of climate change. Cambridge University Press, Cambridge, p 851
Dzioubinski O, Chipman R (1999) Trends in consumption and production: household energy consumption (No. 6). United Nations, Department of Economic and Social Affairs
Ekins P, Speck S (2011) Environmental tax reform (ETR): a policy for green growth. Oxford University Press
Emodi NV (2015) Climate Change in the Nigerian Context. Paper Presentation at the APEC Climate Center, Haeundae-gu Busan, Republic of Korea on 29th June 2015
Emodi VN, Yusuf SD, Boo KJ (2014) The necessity of the development of standards for renewable energy technologies in Nigeria. Smart Grid Renew Energy 5(11):259
Emodi NV, Shagdarsuren G, Tiky AY (2015) Influencing factors promoting technological innovation in renewable energy. Int J Energy Econ Policy 5(3):889–900
Euromonitor International (2014) Consumer Lifestyles in Nigeria. Economic and Consumer. www.euromonitor.com/consumer-lifestyles-in-nigeria/report. Accessed 29 Aug 2015
Geller H (1997) National appliance efficiency standards in the USA: cost-effective federal regulations. Energy Build 26(1):101–109
Gibbs D, O'Neill K (2014) The green economy, sustainability transitions and transition regions: a case study of Boston. Geogr Ann: Ser B Human Geography 96(3):201–216
Glachant JM (ed) (2009) Electricity reform in Europe: Towards a single energy market. Edward Elgar Publishing
Harrison K (2012) A tale of two taxes: the fate of environmental tax reform in Canada. Rev Policy Res 29(3):383–407
Innocent EO (2013) Urban planning problems in Nigeria: a case of Onitsha Metopolis of Anambra state. Singaporean J Bus Econ Manage Stud 1(12)
Intergovernmental Panel on Climate Change. Working Group 3 (2007) Climate Change 2007: Mitigation: Contribution of Working Group III to the Fourth Assessment Report of the Intergovernmental Panel on Climate Change: Summary for Policymakers and Technical Summary. Metz B, Davidson O, Bosch P, Dave R, Meyer L (eds) Cambridge University Press
International Energy Agency (IEA) (2011) World energy outlook 2011. IEA
International Rivers (2008) The World commission on Dams framework: a brief introduction. www.internationalrivers.org/resources/the-world-commission-on-dams-framework-a-brief-introduction-2654. Accessed 24 Aug 2015
Iwayemi A (2008) Nigeria's dual energy problems: policy issues and challenges. Int Assoc Energy Econ 53:17–21
Jaccard M, Agbenmabiese L, Azar C, de Oliveira A, Fischer C, Fisher B, … Zhang X (2012) Global energy assessment: toward a sustainable future
Jacobs M (2012) Green growth: economic theory and political discourse. Centre for Climate Change Economics and Policy Working Paper No, 108, London
Johansson TB, Nakićenović N (eds) (2012) Global energy assessment: toward a sustainable future. Cambridge University Press
Jordán R, Infante B (2012) A strategic planning approach for developing eco-efficient and socially inclusive urban infrastructure. Local Environ 17(5):533–544
Joskow PL (2000) Deregulation and regulatory reform in the US electric power sector. Deregulation of network industries: what's next, pp 113–188

Kim H, Shin ES, Chung WJ (2011) Energy demand and supply, energy policies, and energy security in the Republic of Korea. Energy Policy 39(11):6882–6897
Lamidi WA, Akande LO (2013) A study of status, challenges and prospects of agricultural mechanization in Osun State. J Educ Arts Humanit
Liang QM, Fan Y, Wei YM (2007) Carbon taxation policy in China: how to protect energy-and trade-intensive sectors? J Policy Model 29(2):311–333
Linnér BO, Selin H (2013) The United Nations conference on sustainable development: forty years in the making. Environ Plann C: Govern Policy 31(6):971–987
Majid A, Hussaini IU (2011) The challenges of energy efficiency practice in the Nigerian households
Mathews JA (2012) Green growth strategies—Korean initiatives. Futures 44(8):761–769
Moon TH (2010) Green growth policy in the Republic of Korea: its promise and pitfalls. Korea Observer 41(3):379
National Energy Master Plan (NEMP) (Draft Revised Edition) (2014) Energy Commission of Nigeria (ECN). www.energy.gov.ng. Accessed 20 July 2015
National Energy Policy; Draft Revised Edition (NEP) (2013) Energy Commission of Nigeria (ECN). Federal Republic of Nigeria, Abuja. www.energy.gov.ng. Accessed 6 Aug 2015
National Renewable Energy Laboratory (NREL) (2015) Leading clean energy innovation: renewable portfolio standards. www.nrel.gov/tech_deployment/state_local_governments/basics_portfolio_standards. Accessed 4 Sept 2015
Newscientist (2001) Hydroelectric Power's Dirty Secret Revealed. www.newscientist.com
Osnos E (2011) Faust, China, and Nuclear Power. The New Yorker. www.newyorker.com/news/letter-from-china/faust-china-and-nuclear-power. Accessed 24 Aug 2015
Policy U (2014) Guide to cost-benefit analysis of investment projects. Policy 2014:2020
Presidential Task Force on Power (PTFP) (2013) 2013 Year in Review. The Presidency, Federal Republic of Nigeria. www.africacheck.org/wp-content/uploads/2014/11/PTFP-2013-Year-in-Review-Report-1.pdf. Accessed 23 Aug 2015
Reza Farzin M, Guillaume DM, Zytek R (2011) Iran-the chronicles of the subsidy reform. IMF Working Papers, pp 1–28
Robertson C, Ndebele N, Mhango Y (2011) A survey of the Nigerian middle class. Renaissance Capital, London. www.fastestbillion.com/res/Research/Survey_Nigerian_middle_class-260911.pdf. Accessed 29 Aug 2015
Romero DJ, Castro EO (2008) Mexican energy reform. Law Bus Rev Am 14:859
Sambo AS (2008) Matching electricity supply with demand in Nigeria. International Association for Energy Economics (Fourth Quarter), pp 32–38
Santisirisomboon J, Limmeechokchai B, Chungpaibulpatana S (2003) Least cost electricity generation options based on environmental impact abatement. Environ Sci Policy 6(6):533–541
Sardou IG, Ameli MT, Sepasian MS, Ahmadian M (2013) A novel genetic-based optimization for transmission constrained generation expansion planning. Int J Intell Syst Appl 6(1):73
Shukla PR, Dhar S (2011) Climate agreements and India: aligning options and opportunities on a new track. Int Environ Agreements: Politics Law Econ 11(3):229–243
Sims REH, Schock RN, Adegbululgbe A, Fenhann J, Konstantinaviciute I, Moomaw W, Nimir HB, Schlamadinger B, Torres-Martínez J, Turner C, Uchiyama Y, Vuori SJV, Wamukonya N, Zhang X (2007) Energy supply. In: Metz B, Davidson OR, Bosch PR, Dave R, Meyer LA (eds) Climate Change 2007: Mitigation. Contribution of Working Group III to the Fourth Assessment Report of the Intergovernmental Panel on Climate Change, Cambridge University Press, Cambridge, United Kingdom and New York
Spellman FR (2015) Economics for environmental professionals. CRC Press
Stern N (2007) The economics of climate change: the stern review. Cambridge University press
Sustainable Development Knowledge Platform (SDKP) (2008) Green growth. https://sustainable development.un.org/index.php?menu=1447. Accessed 2 Sept 2015
United Nations (1997) Statistical Division, United Nations. Department for Economic, Social Information, and Policy Analysis. Glossary of environment statistics (No. 67). United Nations Pubns

U.S. Environmental Protection Agency (EPA) (2015) Global greenhouse gas emissions data. www3.epa.gov/climatechange/ghgemissions/global.html. Accessed 1 Sept 2015

Usman ZG, Abbasoglu S (2014) An overview of power sector laws, policies and reforms in Nigeria. Asian Trans Eng 4(2)

Visit Seoul (2015) Metro lines in Seoul metropolitan area. www.visitseoul.net. Accessed 9 Sept 2015

Wang JN, Cai BF, Liu LC, Cao D (2010) Research and practice of low carbon society in China. http://www.caep.org.cn/english/paper/Research-and-Practice-of-Low-Carbon-Society-in-China-20100827.pdf. Accessed 2 Sept 2015

WISE Uranium Project (2015). New Uranium Mining Projects: Africa. www.wise-uranium.org/upafr.html. Accessed 24 Aug 2015

World Bank Databank (2015a) Nigeria: GDP (Current US Dollar). www.databank.worldbank.org/data//reports.aspx. Accessed 3 Sept 2015

World Bank Databank (2015b) Nigeria: Unemployment, total (% of total labor force) (modeled ILO estimate). www.databank.worldbank.org/data//reports.aspx. Accessed 4 Sept 2015

World commission on Dams (WCD) (2000) World commission on Dams Report. www.dams.org. Accessed 23 Aug 2015

World Watch Institute (WWI) (2013) Use and capacity of Global Hydropower Increases. www.worldwatch.org/node/9527. Accessed 23 Aug 2015

Yusuf AA (2008) The Distributional impact of environmental policy: the case of carbon tax and energy pricing reform in Indonesia. Singapore: environment and Economy Program for Southeast Asia, Research report (2008-RR1)

Chapter 6
Conclusion and Policy Recommendations

Abstract This chapter concludes the study with a brief summary of the results of the scenarios and analyses. Based on the implications derived from the scenarios analyzed, cost-benefit analysis and least-cost of electricity generations, some policy recommendations were presented. The chapter also presents limitations from the study and offer suggestions for future research.

Keywords Policy recommendation for sustainable development • Government incentives • Efficiency improvement • Alternative fuels

6.1 Conclusion

The main objective of this study was to develop an energy model for Nigeria that will consider vital factors capable of influencing energy policies in the future, explore the least-cost options for electricity generation and recommend strategic energy policy options for a low carbon development in Nigeria. These objectives were intended to answer the research questions developed from the research problems.

Before developing the model, an extensive review was carried out on the Nigerian energy sector in order to provide an insight into the country's energy resource potentials, energy supply and consumption, and various government agencies and policies. Further review was carried out on the available literature in order to explore the available energy models and literature applying the LEAP model. This provided a literature background and a basis for the selection of the LEAP model for Nigeria. The model developed incorporated four policy scenarios which differ from each other and this was intended to capture the vital factors that may influence energy policies in the future. These factors taken as parameters included GDP, households, population and urbanization growth rates, and growth rates of energy-intensive sectors.

The four scenarios developed are the reference (Ref), low-carbon moderate (LCM), low-carbon advance (LCA) and green optimistic (GO) scenarios.

N.V. Emodi, *Energy Policies for Sustainable Development Strategies*,
Frontiers in African Business Research, DOI 10.1007/978-981-10-0974-7_6

The results of the modeled scenarios showed that energy demand is expected to grow at an annual growth rate of 3.58 % (Ref), 3.53 % (LCM), 2.95 % (LCA), and 2.61 % (GO). The Ref scenario energy demand was the highest with 3,075PJ by 2040, while the GO scenario was the lowest at 2,249.2PJ. The GHG emission rate was very low for GO scenario (124.4 MMTCDE) as compared to other scenarios and this was due to the high level of renewable energy application into the energy mix.

Energy policies such as energy efficiency, fuel and technology switching, were all increased from the LCM scenario which had a moderate policy intervention, the advance LCA scenario, and GO scenario with more aggressive policy intervention. Furthermore, a cost-benefit analysis was carried out in order to ascertain the cost of implementing some policies and strategies in Nigeria which includes energy efficiency and fuel/technology switching. The results showed that it would cost Nigeria 1.69 billion US dollars to implement policies in the LCM, 23.8 billion US dollars for the LCA, and 41.4 billion US dollars in the GO scenario. These investment costs were understandably high, but the offset cost in the long run will come from energy savings on the demand-side, production, and environmental externalities.

The least-cost electricity generation of the power plants proposed in the scenarios showed that among the power plants, on-shore wind power and small hydropower provided the overall least-cost of electricity generation. For fossil fuel power plants, CCGT was identified to be the last option for electricity generation, as well as the lowest GHG emitting power plants besides biomass which was considered a low carbon/renewable energy technology. From the results in general, it was observed that low carbon and renewable technologies will have an important role to play in the realization of a low carbon development in Nigeria.

To achieve this feat, this study explored some strategies that can ensure a low carbon development in Nigeria with a view of attaining green growth. These strategies include thinking and adopting the green growth ideology, energy policy reforms, long-term energy planning and target, energy regulation and standards, environmental tax reforms, urban planning, efficient building design, energy efficiency principle and transport system. The findings from this study can be used as a guide in the development of energy policies and strategies for the attainment of a low carbon development in the long term.

6.2 Policy Recommendations

Based on the policy implications derived from this study, some policy recommendations are provided as follows:

- Provision of domestic water heater and electricity from renewable resources such as solar thermal and PV systems should be deployed to households in Nigeria so as to reduce energy demand and GHGs. This is due to the impact of solar PV system in the household sector in the Nigerian LEAP model which was increased in the alternative scenario. The household sector in GO scenario had a decrease in energy consumption due to the energy efficiency improvement and renewable energy application (solar). This has the potential not only reduce electricity demand, but can be a source of income generation for the consumers (prosumers). This was also highlighted in Sect. 5.3.

- Some form of government incentives such as investment tax credits and low-cost loans should be provided to renewable energy technologies for household, while tax breaks can be offered to electricity companies and industries switching to low-carbon technologies. This was part of the sustainable strategies discussed in Sect. 5.3.3 as part of the long-term energy planning and target. As shown in the cost-benefit analysis in Sect. 5.2.5, funding renewables in the Nigerian context will be expensive in the short term, but this has potential to provide the least-cost option for electricity generation in the long-term as discussed in Sect. 5.2.6.
- Efficiency improvement at both the supply and demand side through modern technologies should be the topmost priority of the government. This was supported in this study as the alternative scenarios proved that the increased impact of energy efficiency has an important role to play in Nigeria's future energy system. The scenario analysis showed that with the increase in energy efficiency policies in the GO scenario, about 826 PJ of energy was saved as against the Ref scenario. On the supply side, energy efficiency improvement in the electricity sector through the reduction in T&D losses contributed to increased electricity supply. Further, investing in efficient power plants such as CCGT should also be considered, because the results of the least-cost electricity generation showed that CCGT power plants were capable of meeting a given electricity demand at a least cost and least GHG emission. Therefore, with the increased implementation of energy efficiency policies in Nigeria, more energy and GHG emission will be reduced.
- Alternative fuels in the transport sector, such as biofuels and CNG will aid in the reduction in oil dependency and GHGs emission. Stronger promotion of these alternative fuels should be encouraged by the government. This is due to the results obtained from the alternative scenarios. In the GO scenario, LPG, CNG, and biofuels options were made available for the transport sector and this did not only reduce GHG emissions, but also reduced the dependency on gasoline and diesel.
- Nigeria's energy policy should be reviewed at a shorter time frame with the aid of policy evaluation experts who can ascertain the impact of government policies at the grassroot level. This impact is presented in Fig. 2.28 where some policy overlaps in the Nigerian energy polices were highlighted. In order to implement the policies in the alternative scenarios, policy evaluation body should be established. In the Nigerian LEAP model, residential households were categorized into urban and rural households. Most government policies are less effective in the rural areas, while those that have impact on the urban areas are sometimes left obsolete. Therefore, energy polices need to be reviewed

at a shorter time frame, while policy evaluation units should also estimate the impact of government policies at the grassroot level.

- The setting up of an aggressive environmental tax by relevant government agencies will aid in the reduction of GHG emissions as the society will become more aware of the effect of GHGs accumulation. Although this was not part of the scenario analysis, it was discussed as sustainable strategy that can be employed by the Nigerian government as discussed in Sect. 5.3.5. This can be an energy reduction measure as a shift will be made from tax base to resource consumption. The funds realized can also be used to fund the development of renewable energy through the provision of soft loans and incentives.
- The government through its relevant bodies and agencies should ensure the development of proper urban planning to accommodate energy efficiency in buildings, while providing a walkable way. Some challenges may occur in order to fulfil this policy. However, incentives have the ability to stimulate consumers, house owners, and building contractors. As the cost-benefit analysis showed in Sect. 5.2.5, the government should look into the long-term benefit of energy efficiency in buildings as this could help in saving the cost of delivering future energy services.
- The increase in private car ownership should be discouraged by the government as this would increase energy demand and GHG emissions. This can be achieved through the provision of efficient public transport system and tax payments for private cars ownership and parking lots for private cars. The four policy scenarios showed an increased amount of cars from the base year (2010), to 2040. Although fuel options and energy efficiency measures were in place to reduce energy demand and GHG emissions, the rate of private cars was still high. This calls for increased measures to see to its reduction as discussed in Sect. 5.3.6.
- Besides the need to increase the production capacity of refineries in Nigeria, the government needs to introduce, deploy, and promote the use of portable crude oil refineries as a short- and medium-term policy. This will address the nation's current and future energy needs, while expansion of the domestic refineries continues as a long-term policy plan. The energy balance under the Ref scenario (Sect. 5.1.1.3) gave an insight on how the Nigerian energy system will be by 2040 in the case of high importation of oil products. The production capacity of the refineries in Nigeria cannot possibly meet future energy requirements. Thus, the government needs to devise other means of supporting the demand for fossil fuel products. The construction of large refinery facilities is considerably high and requires long-term planning in order to be fully operational. This supports the recommendation for the introduction of portable refineries which are relatively cheaper and more efficient to meet short-term demands. Stakeholders in the oil and gas sector need to be encouraged by the Nigerian government to key-into the idea of portable refineries as short- to medium-term strategies.

6.3 Limitations and Suggestions for Further Study

This study is not without limitations and this is due to the nature of the research methodology applied. The limitations of this study include:

- The incomplete variables in the transport sector which include railway, airway, and seaway transport systems. The datasets for these variables were not available during the development of the Nigerian LEAP model. Therefore, the provision of these datasets will boost the results of the current model.
- The four scenarios developed in this study do not exhaust the possibility of providing solutions to a low-carbon development in Nigeria. The future is uncertain, hence more pathways that involve various energy policies could be developed in further studies and the results compared to identify a better pathway.
- The study did not observe some emission reduction schemes that can earn some revenues such as the clean development mechanisms (CDM) among others. This will also provide more sustainable options for a low-carbon development in Nigeria if integrated into the Nigerian LEAP model.
- The impact of fuel subsidy removal and oil price change was not taken into consideration in the development of policy scenarios. This can be added to one or more scenarios in further studies so as to identify the adoption measures for Nigeria in the future.
- Finally and most importantly, the Nigerian LEAP model did not consider the impact of climate change on the country's future energy system. This is important because power plants such as hydropower plants are highly prone to droughts, and thermal gas power plants need water to cool their turbines. Also the identification of possible wind and solar locations with high wind speed and solar radiations is vital in energy planning.

In conclusion, there exists room for more improvement in this study if the limitations listed out are addressed in future studies. However, this study expects that the strategies and recommendations which were based on the analysis carried out on the Nigerian LEAP model will indeed move Nigeria towards a low-carbon society that has the mindset of attaining green growth in the future.

Appendix A
Projections

N.V. Emodi, *Energy Policies for Sustainable Development Strategies*,
Frontiers in African Business Research, DOI 10.1007/978-981-10-0974-7

Table A.1 Projected electricity generation capacity under the reference scenario (2010–2040)

	2010		2015		2020		2025		2030		2035		2040	
	MW	Share (%)	MW	Share (%)	MW	Share (%)	MW	Share (%)	MW	Share (%)	MW	Share (%)	MW	Share (%)
On-grid electricity generation technologies														
Biomass					3	0	16	0	25	0	35	0	54	0
CCGT	1100	12	2000	9	3500	10	5000	11	15000	21	21000	22	26300	24
Coal Steam					1000	3	3500	7	5000	7	10000	11	13000	8
Hydropower	1900	20	2100	9	5000	14	6500	14	7000	10	7500	8	9000	8
Nuclear					1000	3	1500	3	2500	4	3500	4	4000	4
On-shore wind					10	0	19	0	22	0	25	0	29	0
SCGT	6500	68	18000	81	25000	70	30000	64	40000	57	52000	55	55260	51
Solar thermal					1	0	10	0	20	0	30	0	40	0
Solar PV			30	0	75	0	200	0	425	1	560	1	600	1
Subtotal	9500		22130		35589		46745		69992		94650		108283	
Off-grid electricity generation technologies														
Diesel gen	3000	70	4600	54	6500	43	9600	44	13000	46	19000	50	24000	51
Gasoline gen	1300	30	2600	31	3800	25	5000	23	5500	19	6000	16	7000	15
Gas turbine			1300	15	4900	32	7000	32	10000	35	13000	34	16000	34
Subtotal	4300		8500		15200		21600		28500		38000		47000	
Total	13800		30630		50789		68345		98492		132650		155283	

Table A.2 Projected share of vehicles by fuel type in the LCM scenario

Vehicle type	Fuel type	Share in 2010 (%)	Share in 2040 (%)
Motorcycles	Petrol	100	80
	Biodiesel	0	20
Cars	Petrol	99	15
	Diesel	1	5
	Biofuel	0	70
Light goods vehicle (LGV)	Petrol	100	50
	Biofuel	0	50
Urban bus (UB)	Petrol	100	30
	Biofuel	0	70
Long-distance coach (LDC)	Diesel	100	40
	Biofuel	0	60
Heavy goods vehicle (HGV)	Diesel	100	40
	Biodiesel	0	60

Table A.3 Charcoal production shares by scenarios

Scenario	Production plants	Share (%)
Base year	Traditional earth mound	100
Reference	Traditional earth mound	70
	Brick kiln	30
Low carbon moderate	Traditional earth mound	30
	Steel kiln	40
	Casamance	30
Low carbon advance	Traditional earth mound	25
	Mud beehive kiln	40
	Adam retort kiln	35
Green optimistic	Traditional earth mound	20
	Single drum kiln	40
	Brick beehive kiln	40

Table A.4 Projected electricity generation capacity under the LCM Scenario (2010–2040)

	2010		2015		2020		2025		2030		2035		2040	
	MW	Share (%)	MW	Share (%)	MW	Share (%)	MW	Share (%)	MW	Share (%)	MW	Share (%)	MW	Share (%)
On-grid electricity generation technologies														
Biomass					400	1	800	2	1000	2	1500	2	2000	2
CCGT	1100	12	9000	37	13000	38	15000	33	17000	29	19000	24	25000	22
Coal Steam					2000	6	3000	7	6000	10	8000	10	14000	12
Coal supercritical							1500	3	4000	7	7000	9	11000	10
Hydropower	1900	20	3000	12	4000	12	5500	12	7000	12	8500	11	10000	9
Nuclear					1000	3	2000	4	3000	5	4000	5	5000	4
On-shore wind			20	0	30	0	40	0	50	0	60	0	100	0
SCGT	6500	68	12000	50	14000	40	16000	35	17000	29	24000	30	35000	31
Solar thermal					100	0	200	0	500	1	800	1	1000	1
Solar PV					100	0	2000	4	4000	7	7000	9	10000	9
Subtotal	9500		24020		34630		46040		59550		79860		113100	
Off-grid electricity generation technologies														
Diesel gen	3000	70	4600	54	6000	38	9000	35	10000	31	15000	33	17000	30
Gasoline gen	1300	30	2600	31	4500	28	6000	24	8500	27	11000	24	14000	24
Gas turbine			1300	15	3500	22	7800	31	9500	30	14000	30	18000	31
Small hydropower					1600	10	2000	8	2500	8	3000	7	3400	6
Solar PV					200	1	600	2	1500	5	3000	7	5000	9
Subtotal	4300		8500		15800		25400		32000		46000		57400	
Total	13800		32520		50430		71440		91550		125860		170500	

Table A.5 Projected share of vehicles by fuel type in the LCA scenario

Vehicle type	Fuel type	Share in 2010 (%)	Share in 2040 (%)
Motorcycles	Petrol	100	60
	Biodiesel	0	40
Cars	Petrol	99	20
	Diesel	1	0
	Biofuel	0	40
	LPG	0	40
Light goods vehicle (LGV)	Petrol	100	30
	Biofuel	0	30
	LPG	0	40
Urban bus (UB)	Petrol	100	30
	Biofuel	0	40
	LPG	0	30
Long-distance coach (LDC)	Diesel	100	25
	Biofuel	0	40
	LPG	0	35
Heavy goods vehicle (HGV)	Diesel	100	30
	Biodiesel	0	70

Table A.6 Projected electricity generation capacity under the LCA scenario (2010–2040)

	2010		2015		2020		2025		2030		2035		2040	
	MW	Share (%)	MW	Share (%)	MW	Share (%)	MW	Share (%)	MW	Share (%)	MW	Share (%)	MW	Share (%)
On-grid electricity generation technologies														
Biomass					500	2	2000	4	4000	6	5000	5	7500	5
CCGT	1100	12	6000	35	9000	32	14000	29	19000	27	25000	25	30000	22
Coal CFB			1000	6	3000	11	5000	10	7000	10	10000	10	14000	10
Coal IGCC					1000	4	4000	8	7000	10	9000	9	13000	10
Hydropower	1900	20	3000	18	5000	18	7000	14	9000	13	11000	11	14000	10
Nuclear					1000	4	3000	6	5000	7	9000	9	15000	11
On-shore wind			20	0	100	0	1500	3	3000	4	5000	5	8000	6
SCGT	6500	68	7000	41	8000	28	9000	18	10000	14	11000	11	12000	9
Solar thermal					300	1	800	2	2500	4	5500	6	9000	7
Solar PV					500	2	2500	5	4000	6	8000	8	14000	10
Subtotal	9500		17020		28400		48800		70500		98500		136500	
Off-grid electricity generation technologies														
Diesel gen	3000	70												
Gasoline gen	1300	30												
Gas turbine			1000	40	3000	31	5000	30	7000	27	9000	26	11000	25
SCGT			1500	60	3500	36	5500	33	8000	30	10000	28	12000	27
On-shore wind					500	5	1800	11	3000	11	4500	13	6500	15
Small hydropower					2000	21	3500	21	4500	17	6000	17	7500	17
Solar PV					600	6	1000	6	3800	14	5600	16	7000	16
Subtotal	4300		2500		9600		16800		26300		35100		44000	
Total	13800		19520		38000		65600		96800		133600		180500	

Table A.7 Projected share of vehicles by fuel type in the GO scenario

Vehicle type	Fuel type	Share in 2010 (%)	Share in 2040 (%)
Motorcycles	Petrol	100	50
	Biodiesel	0	50
Cars	Petrol	99	20
	Diesel	1	0
	Biofuel	0	30
	LPG	0	25
	CNG	0	25
Light goods vehicle (LGV)	Petrol	100	10
	Biofuel	0	20
	LPG	0	35
	CNG	0	35
Urban bus (UB)	Petrol	100	5
	Biofuel	0	25
	LPG	0	35
	CNG	0	35
Long-distance coach (LDC)	Diesel	100	30
	Biofuel	0	20
	LPG	0	25
	CNG	0	25
Heavy goods vehicle (HGV)	Diesel	100	20
	Biodiesel	0	45
	CNG	0	35

Table A.8 Projected electricity generation capacity under the GO scenario (2010–2040)

	2010		2015		2020		2025		2030		2035		2040	
	MW	Share (%)	MW	Share (%)	MW	Share (%)	MW	Share (%)	MW	Share (%)	MW	Share (%)	MW	Share (%)
On-grid electricity generation technologies														
Biomass					800	2	2500	4	4500	5	6500	5	8000	5
CCGT	1100	12	6000	36	8000	24	10000	17	12000	13	14000	12	16000	11
Coal CFB			1000	6	2500	7	4000	7	6000	7	7000	6	9500	7
Coal IGCC					1000	3	2000	3	4000	4	6000	5	8000	5
Coal supercritical							1000	2	3000	3	4000	3	5000	3
Geothermal					1000	3	2500	4	4000	4	6000	5	8500	6
Hydropower	1900	20	3000	18	6000	18	8000	13	10000	11	13000	11	15500	11
Nuclear					2000	6	6000	10	10000	11	14000	12	18000	12
Off-shore wind					100	0	2000	3	5000	6	7000	6	9000	6
On-shore wind			20	0	500	1	4000	7	7000	8	10000	8	12000	8
SCGT	6500	68	6700	40	7000	21	7300	12	7800	9	8000	7	8500	6
Solar thermal					2000	6	5000	8	7000	8	9000	8	11000	8
Solar PV					3000	9	6000	10	9000	10	15000	13	17000	12
Subtotal	9500		16720		33900		60300		89300		119500		146000	
Off-grid electricity generation technologies														
Biomass					500	9	2000	16	3500	17	4500	16	6000	17
Diesel gen	3000	70												
Gasoline gen	1300	30												
SCGT			1500	100	1700	30	2000	16	2500	12	3300	12	4000	11
On-shore wind					500	9	2000	16	4000	20	6000	22	8000	23
Small hydropower					2000	35	3500	28	5500	27	6800	25	8000	23
Solar PV					1000	18	3000	24	5000	24	7000	25	9000	26
Subtotal	4300		1500		5700		12500		20500		27600		35000	
Total	13800				39600		72800		109800		147100		181000	

Appendix B
Datasets Used in the Development of Nigeria LEAP Model

N.V. Emodi, *Energy Policies for Sustainable Development Strategies*,
Frontiers in African Business Research, DOI 10.1007/978-981-10-0974-7

Table B.1 Socioeconomic data

Data type	Value in 2010
GDP	369 million USD
GDP growth rate	8 %
Income	2310 USD
Income growth rate	4.9 %
Population	160 million
Population growth rate	2.55 %
Household	37 million
Household size	5 People
Household growth rate	3.16 %
Urbanization growth rate	44 %
Urban population with electricity	79.80 %
Urban population without electricity	20.2 %
Rural population with electricity	34.90 %
Rural population without electricity	65.1 %
Access to electricity	48 %
Household source of electricity supply PHCN only	67.40 %
Household source of electricity supply from PHCN/Generator	27.60 %
Household source of electricity supply from private generator only	1.90 %
Household source of electricity supply from rural electrification/generator	2.20 %
Household source of electricity supply from rural electrification only	0.80 %
Household source of electricity supply from solar PV	0.10 %

Source The World Bank. www.data.worldbank.org
National Bureau of Statistics. www.nigerianstat.gov.ng
Index Mundi: www.indexmundi.com/facts/nigeria/access-to-electricity

Table B.2 Percentage distribution of household source of fuel for cooking (2010)

Distribution of household source of fuel for cooking	Percentage
Firewood	69.8
Charcoal	0.84
Kerosene	26.6
Gas	1.11
Electricity	0.52
Sawdust/plant residue	0.09
Animal waste	0.07
Others	0.84

Table B.3 Percentage distribution of household by source of fuel for lighting (2010)

Distribution of household by source of fuel for lighting	Percentage
Kerosene	31.70
Gas	0.1
Electricity	64.80
Generator	3
Solar PV	0.40

Table B.4 Contributions to GDP by percentage share and growth rate (2010)

Contributions to GDP (2010)	Percentage share (%)	Growth rate (%)
Agriculture	39.21	3.97
Solid mineral	0.38	12.52
Crude petroleum and natural gas	13.76	0.91
Manufacturing	4.20	7.55
Telecommunication	7.05	31.83
Finance and insurance	3.37	4.05
Wholesale and rental trade	19.92	9.61
Building and construction	2.19	12.58
Hotel and restaurant	0.55	12.15
Real estate	1.85	10.41
Business and other services	0.95	9.69
Others	6.58	5.18

Source National Bureau of Statistics. www.nigerianstat.gov.ng

Table B.5 Petroleum product import (2010)

Petroleum product import		Value
Premium motor spirit (PMS)	Quality (metric tons)	649,896.40
	Value ($)	702,249,321.63
Dual-purpose kerosene (DPK)	Quality (metric tons)	774,227.20
	Value ($)	837,190,427.87

Source National Bureau of Statistics. www.nigerianstat.gov.ng
Nigeria National Petroleum Corporation. www.nnpcgroup.com

Table B.6 Petroleum product export (2010)

Petroleum product export		Value
High pour fuel oil (HPFO)	Quality ('000 L)	211,566.69
	Value ($)	115,733.15
Low pour fuel oil (LPFO)	Quality ('000 L)	543,090.98
	Value ($)	300,147.03
Naphtha	Quality ('000 L)	614,476.50
	Value ($)	293,158.06
Natural gas liquids (NGL)/liquefied petroleum gas (LPG)	Quality (BBLS)	697,428
	Value ($)	537,623,370.63
Nigeria liquefied natural gas (NLNG)	Quality (MMBTU)	727,962,414
	Value ($)	1,665,477,338.74
Crude oil	Quality (MMBTU)	822,082,224
	Value ($)	92,119,954,888.10

Source National Bureau of Statistics. www.nigerianstat.gov.ng
Nigeria National Petroleum Corporation. www.nnpcgroup.com

Table B.7 Kaduna refining and petrochemical company actual quality of output in 2010

Refining and petrochemical product	Actual quantity of output
Premium motor spirit (PMS) (i.e., Petrol)	210,172
Dual-purpose kerosene (DPK)	120,826
Automotive gas oil (AGO) (i.e., Diesel)	250,947
Residual fuel oil	237,655
Asphalt	25,292
Fuel and losses	193,151
Total	1,038,043

Source National Bureau of Statistics. www.nigerianstat.gov.ng
Nigeria National Petroleum Corporation. www.nnpcgroup.com

Table B.8 Nigeria energy balance (2010) (KTOE)

	Coal and peat	Crude oil	Oil products	Natural gas	Hydro	Biofuels and waste	Electricity	Total
Production	23	132,372	0	26,573	548	97,809	0	257,326
Imports	0	0	9,134	0	0	0	0	9134
Exports	0	−128,106	−905	−17750	0	0	0	−146,761
International marine bunkers	0	0	−413	0	0	0	0	−413
International aviation bunkers	0	0	−174	0	0	0	0	−174
Stock changes	0	773	85	0	0	0	0	858
Total primary energy supply (TPES)	23	5,039	7727	8,823	548	97,809	0	119,970
TPES (%)	0.02	4.20	6.44	7.35	0.46	81.53	0	100
Sectors								
Industry	23	0	264	607	0	5919	279	7093
Transport	0	0	9414	0	0	0	0	9414
Household	0	0	2281	0	0	83675	1029	85287
Commercial/services	0	0	6			1532	469	2007
Agriculture	0		10	0	0	0	0	10

Source International Energy Agency (IEA) 2015. http://www.iea.org/statistics/statisticssearch/report/?year=2010&country=NIGERIA&product=Balances

Table B.9 Value added and growth rate of sectors (2010)

Sector	Value added	Growth rate
Industry	80,062,248,999	6 %
Commercial and service	196,477,501,364	12 %
Agriculture	86,820,135,996	6

Source World Bank. www.data.worldbank.org

Table B.10 Coal mining in Nigeria (2010)

Type of coal	2010 Production	Production efficiency
Sub-bituminous	41.89 thousand tons of coal equivalent	50

Source Nigeria Coal Production by Year. U.S. Energy Information Administration (EIA). Available online: www.eia.gov

Table B.11 Distribution of Nigeria average household appliances (2010)

Appliances	Currently owned, out of 100 (%)
Air-conditioning	47
Refrigerator	87
Fridge freezer (food preservation)	42
Separate deep freezer	30
Electric kettle (water heating)	70
Electric fan	85
Hi-Fi/stereo system	53
Video player	33
DVD player	94
TV	90

Source A survey of Nigerian middle class. http://www.fastestbillion.com/res/Research/Survey_Nigerian_middle_class-260911.pdf

Table B.12 Energy intensities in Nigerian households (2010)

Usage	KW/h[a]	Share (%)
Ventilation and air-conditioning	191.04	12
Lighting	506.28	32
Refrigeration	402.01	25
Cooking (electric)	46.44	3
Food preservation	47.64	3
Water heating	0.6	0
Electronic (TV, HI-Fi, Electric fan)	395.04	25
Total	1589.76	100

[a]Converted values
Source Irimiya et al. (2013)
National Bureau of Statistics. www.nigerianstat.gov.ng
Energy Commission of Nigeria. www.energy.gov.ng
Ibitoye (2013)

Table B.13 Percentage share by area for lighting (Fuel) (2010)

Area	Electricity	LPG	Kerosene	Others	Total (%)
Urban (electrified)	46	8.3	39	5.7	100
Urban (non-electrified)	15	25	50	10	100
Rural (electrified)	18	–	70	12	100
Rural (non-electrified)	5	–	81	14	100

Table B.14 Percentage share of area for cooking (Fuel) (2010)

Area	Electricity	LPG	Kerosene	Firewood	Charcoal	Sawdust/ plant residue	Total (%)
Urban (electrified)	24	33	42	1	–	–	100
Urban (non-electrified)	23	32	43	3	–	–	100
Rural (electrified)	9	3	53	29	5	1	100
Rural (non-electrified)	–	–	48	44	6	4	100

Table B.15 Categories of major vehicles in Nigeria (2010)

Type	Classification	Fuel system	Fuel type	Type of CC	Share (%)	Average age (year)
Motorcycle	Scooters	2-stroke	Petrol	>50 cc	40	5
	Motorcycles	2-stroke	Petrol	>50 cc	50	
	Motorcycles	4-stroke	Petrol	<250 cc	5	
	Motorcycles	4-stroke	Petrol	<240 cc	5	
Cars	Lower-medium	Diesel pump	Diesel	<2.0L	1	14
	Lower-medium	Carburetor	Petrol	<2.0L	15	
	Lower-medium	Fuel injection	Petrol	<2.0L	25	
	Upper-medium	Carburetor	Petrol	<2.0L	15	
	Upper-medium	Fuel injection	Petrol	<2.0L	15	
	SUV	Carburetor	Petrol	<1.4L	15	
	SUV	Carburetor	Petrol	<1.4L	14	
Light goods vehicle (LGV)	Minivan/truck	Carburetor	Petrol	<3.5t	75	24
	Minivan/truck	Fuel injection	Petrol	<3.5t	25	
Urban bus (UB)	Light bus	Diesel pump	petrol	<3.5t	50	16
Long-distance Coach (LDC)	Medium bus	Diesel pump	Diesel	16t	50	
Heavy goods vehicle (HGV)	Light truck	Diesel pump	Diesel	5t	17	18
	Medium truck	Diesel pump	Diesel	16.2t	12	
	Heavy truck	Diesel pump	Diesel	25t	31	
	Trailer	Diesel pump	Diesel	40t	40	

Source Babatunde and Akintayo (2012)

Table B.16 Vehicle fleet, activity and fuel efficiency in Nigeria (km/L) (2010)

Type	Fleet	Veh/km	km/L (Petrol)	km/L (Diesel)
Motorcycle	3,322,888	23,260	40	
Car	4,650,509	79,059	11	10
LGV	557,301	16,719	–	10
UB	130,089	160	20	–
LDC	5,344	240	–	3.5
HGV	130,089	240	–	3

Source Cervigni et al. (2013)
International Association of Public Transport (UITP, from the French: L'Union internationale des transports publics) and African Association of Public Transport (UATP, from the French: L'Union Afrique des transports publics) (2010). Report on Statistical Indicators of Public Transport Performance in Africa. Available online: www.uitp.org

Table B.17 Charcoal production plants and conversion efficiencies (2010)

Production plants	Conversion efficiency (%)
Traditional earth mounds	20
Single drum (SD) kilns	44
Mud beehive (MBH) kilns	43
Brick beehive (BBH) kilns	47
Adam retort	40
Casamance kilns	30
Brick kilns	30
Steel kilns	35

Source Energypedia. Available online: www.energypedia.info/wiki/Charcoal_Production
Bioenergy Lists. Available online: www.bioenergylists.org/stovesdoc/Smith/kilns.htm

Table B.18 Nigeria crude oil refineries (2010)

Refinery	Location	Barrels per day (bbl/d)	Barrels per year (bbl/y)	Efficiency (%)	Facility capacity (Mt/year)
Kaduna	Kaduna state	110,000	40,150,000	90	6 million
Port Harcourt	Rivers state	210,000	76,650,000	60	11 million
Warri	Delta state	125,000	45,625,000	80	6 million

Source Oil Refining in Nigeria-Overview. MBendi. Available online: http://www.mbendi.com/indy/oilg/ogrf/af/ng/p0005.htm
Oil and Gas. Country Studies. Available online: www.countrystudies.us/nigeria/65.htm
Premium Times. Available online: www.premiumtimesng.com/regional/nwest/187584-kaduna-refinery-to-hit-90-production-efficiency-says-md.html
Premium Times. Available online: www.premiumtimesng.com/business/oil-and-gas/187477-port-harcourt-warri-refineries-commence-production-says-nnpc.html

Table B.19 Conventional energy reserves in Nigeria and their potentials (2010)

Resource type	Reserves		Production	Domestic utilization (natural units)
	Natural units	Energy units (Btoe)		
Natural gas	187 trillion SCF	4.19	6 billion SCF/day	3.4 billion SCF/day
Crude oil	36.22 billion barrels	5.03	2.5 million barrels/day	450,000 barrels/day
Tar sands	31 billion barrels of equivalent	4.31	Insignificant	Insignificant
Coal and lignite	2.175 billion ton	1.52	–	–
Nuclear element	None	–	–	–

Source NBS (National Bureau of Statistics). National account. Abuja: Federal Republic of Nigeria

Table B.20 Renewable energy resources in Nigeria and their potential (2010)

Resource type	Reserves		Production	Domestic utilization (natural units)
	Natural units	Energy units (Btoe)		
Small hydropower	3500 MW	0.34 (over 40 years)	30 MW	30 MW
Large hydropower	11,250 MW	0.8 (over 40 years)	1938 MW	1938 MW
Wind	2–4 m/s at 10 m height (main land)	0.0003 (4 m/s @ 12 % probability, 70 m height, 20 m rotor, 0.1 % land area, 40 years	–	–
Solar radiation	3.5–7.0 kWh/m2/day (4.2 million MWh/day using 0.1 % land area)	5.2 (40 years and 0.1 % land area)	6 MWh/day	6 MWh/day
Biomass fuel wood	11 million hectares of Forest and wood land Excess of 1.2 m ton/day	–	0.120 million ton/day	0.120 million ton/day
Animal waste	211 million assorted animals	–	0.781 million ton of waste/day	None
Energy crops and agricultural residue	28.2 million hectares of arable land (=30 % of total land)	–	0.256 million ton of assorted crops/day	None

Source Energy Commission of Nigeria (ECN). Renewable Energy Master Plan

Table B.21 Cost of primary energy resources (2010)

Primary resources	Cost (2010)
Coal	$61/ton
Natural gas	$4.4/MCF
Crude oil	$80/barrel
Nuclear (Uranium)	$60/lb (i.e. Pound)

Source The Nigerian Petroleum Product Pricing Regulatory Agency (PPPRA). Available online: www.pppra.gov.ng
Coal Price. U.S. Energy Information Administration (EIA). Available online: www.eia.gov/totalenergy/data/annual/showtext.cfm?t=ptb0709
Natural Gas Price: International Monetary Fund Primary Commodity Prices. Available online: www.imf.org/external/np/res/commod/index.aspx
Crude Oil Price. U.S. Energy Information Administration (EIA). Available online: www.eia.gov/dnav/pet/pet_pri_spt_s1_a.htm
Uranium Price. Uranium Info. Available online: www.uranium.info/daily_u3o8_spot_price_indicator.php, www.acap.com.au/uranium-market

Table B.22 Capital cost of cooking technologies in Nigeria (2010)

Type of stove	Capital cost (US$/kw)
Electric	76.2
Kerosene	6.1
LPG	76.2
Biomass	0.6

Source The International Atomic Energy Agency; Energy Master Plan Study for Nigeria (2004). Energy Commission of Nigeria (ECN): Abuja, Nigeria
World Bank. 2004. Nigeria LP Gas Sector Improvement Study. World Bank, Washington, DC. © World Bank. Available online: www.openknowledge.worldbank.org/handle/10986/18859

Table B.23 Cost of secondary energy resources (2010)

Secondary resources	Cost (2010)
Gasoline	$0.41/L
Diesel	$0.88/L
LPG	$0.40/kg
Kerosene	$0.32/L
Lubricant	$3.5/L
Avgas	$1.07/L
CNG	$0.68/L
Residual fuel oil	$1.28/L
Charcoal	$0.03/kg
Firewood	$0.02/kg

Source Price of Gasoline and Diesel in Nigeria. Trading Economics. Available online: www.tradingeconomics.com/nigeria/pump-price-for-gasoline-us-dollar-per-liter-wb-data.html
LPG Price. Global Petrol Prices. Available online: www.globalpetrolprices.com/lpg_prices
Price of kerosene. Sunnewsonline. Available online: www.sunnewsonline.com/new/profiteering-and-the-high-cost-of-kerosene, www.energypedia.info/wiki/Fuel_Prices_Nigeria
Price of lubricant. Jumia. Available online: www.jumia.com.ng/oil-fluid
Price of Avgas in Nigeria. African Manager. Available online: www.africanmanager.com/site_eng/detail_article.php?art_id=12743
Prices of CNG and LPG. My Petrol Price. Available online: www.mypetrolprice.com/2/CNG-price-in-Delhi
Price of Residual Fuel Oil. U.S. Energy Information Administration (EIA). Available online: www.eia.gov/dnav/pet/hist/LeafHandler.ashx?n=PET&s=EMA_EPPR_PWG_NUS_DPG&f=M
Haran Resources. Available online: www.haranresources.com/hardwood-charcoal.html
The International Atomic Energy Agency; Energy Master Plan Study for Nigeria (2004). Energy Commission of Nigeria (ECN): Abuja, Nigeria
World Bank. 2004. Nigeria LP Gas Sector Improvement Study. World Bank, Washington, DC. © World Bank. Available online: www.openknowledge.worldbank.org/handle/10986/18859

Table B.24 Characteristics of household cooking technologies (2010)

Stove type	Number of stoves (1000 s)	Unit capacity (kw)	Efficiency (%)	Stove lifetime (years)	Operating time (hr/year)	Amount of fuel consumed (kt/year)	Energy in Fuel consumed (PJ/year)	Useful energy (PJ/year)
Electricity	23	2.5	95	10	1200	NA	0.26	0.25
Charcoal	648	2	22	5	1500	936	31.9	7.0
LPG	56	2.5	60	10	1200	21	1.0	0.6
Kerosene	2407	2	55	10	1500	1,116	47.3	26.0
Firewood	19,565	2	18	5	1500	109,478	1173.6	211.3
Plant residue	250	2	10	5	1500	2807	27.4	2.7
Total	22,949					114,358	1281.4	247.9

Source International Energy Agency (IEA), 2015. Available online; www.iea.org/statistics/statisticssearch/report/?year=2010&country=NIGERIA&product=Balances
Nigeria Energy Statistics. United Nations (UN) Data Statistics Division. Available online: www.data.un.org/Search.aspx?q=Nigeria
International Institute for Sustainability Analysis and Strategy (IINAS). Global Emission Model for Integrated Systems (GEMIS); Version 4.3; IINAS: Darmstadt, Germany, 2006
Gujba et al. (2015)

Table B.25 Environmental externality cost of greenhouse gases (GHGs)

GHG	Environmental externality cost
Nitrous oxide	4.4 USD/kg
Sulfur dioxide	2 USD/kg
Carbon dioxide non-biogenic	0.22 USD/kg
Carbon monoxide	0.96 USD/kg
Nitrogen oxides	7 USD/kg
Methane	0.24 USD/kg

Source DOE, U. Electricity generation and environmental externalities: Case studies. Report of Energy Information Administration within the US Department of Energy, DOE/EIA, 598

Table B.26 Power plant technology information (2010)

	SCGT	CCGT	Geothermal	Biomass	Large hydro	Small hydro	Nuclear	Solar thermal	Solar PV	Coal CFB	Coal IGCC	Coal SC	Coal steam	On-shore wind	Off-shore wind
Investment ($/kw)	582	612	1000	1721	1014	1980	1946	2309	3923	1000	1380	1050	978	1253	3940
Fixed O&M ($/kw)	11	16	40	50	13	12	100	53	11	39	34	26	8	25	46
Variable O&M ($/kw)	0.003	0.005	0.006	0.008	0.0005	–	0.005	–	–	1.15	1.87	0.75	0.004	0	0
Load Type	Peak	Peak	Base	Base	Base	Base	Base	Base	Base	Base	Base	Base	Base	Base	Base
Process Efficiency (%)	40	58	21	80	59	60	36	23	15	42	45	44	30	27	43
Maximum Availability (%)	90	95	90	83	60	60	90	22	42	80	80	80	80	43	45
Capacity Credit (%)	100	100	80	100	100	100	100	30	30	100	100	100	100	30	30
Life time (Years)	30	30	40	40	80	25	60	25	25	40	40	40	40	25	25

Source OECD I (2010) Projected costs of generating electricity

- Khatib (2010)
- Carl B. Alston Power Plant Economics. Available online: www.wpweb2.tepper.cmu.edu/ceic/SeminarPDFs/Carl_Bozzuto_Seminar.pdf
- Salvatore (2013)
- Breeze (2014)
- Electricity Market Modules. Energy Information Administration (EIA) (2009). Available online: www.eia.gov/oiaf/archive/aeo07/assumption/pdf/electricity.pdf
- Variable Operations and Maintenace (VOM) Costs: Educational Document. Available online: www.pjm.com/~/media/committees-groups/subcommittees/cds/20110321/20110321-item-04b-educational-paper-for-vom.ashx
- Wiltsee (2000)
- Biomass for Power Generation and CHP. International Energy Agency (IEA) Technology Essentials. Available online: www.iea.org/publications/freepublications/publication/essentials3.pdf
- New Biomass and Biogas Generation. Available online: https://ethree.com/downloads/.../14%20Biomass%20Assumptions%20v2
- Resource, Cost, and Performance Assumptions
- Tidball et al. (2010)
- Zarrouk and Moon (2014)
- Taylor et al. (2008)
- Branker et al. (2011)
- Efficiency in Electricity Generation. Union of the Electricity Industry/VGB Power Tech. Available online: www.eurelectric.org/Download/Download.aspx?DocumentID=13549
- Sims et al. (2010)
- Cost and Performance Review of Generation Technologies. Energy and Environmental Economics, Inc. Available online: www.nwcouncil.org/media/6867814/E3_GenCapCostReport_finaldraft.pdf
- Power Plant O&M: how does the industry stack up on cost? Power Technology. Available online: www.power-technology.com/features/featurepower-plant-om-how-does-the-industry-stack-up-on-cost-4417756
- Distributed Generation Renewable Energy Estimate of Costs. Energy Analysis. National Renewable Energy Laboratory (NREL). Available online: www.nrel.gov/analysis/tech_lcoe_re_cost_est.html
- The Economics of Nuclear Power. World Nuclear Association. Available online: http://www.world-nuclear.org/info/Economic-Aspects/Economics-of-Nuclear-Power
- Stefansson (2002)

Table B.27 Electricity generation information (2010)

Power station	Location	Plant type	Historical production (MWh)	Installed capacity (MW)	Average availability (MW)	Availability factor (MW)	No. of units	Year of constriction (Age)	Contribution to the national grid (%)
Thermal									
AES Barge	Egbin	SCGT	1,538,650	270	236.44	0.78	NA	NA	NA
Afam VI-V	Rivers	SCGT	56,223.40	726	63.52	0.07	18	1963	1.8
Afam VI	Rivers	CCGT	2,927,275.02	624	322.82	0.65	NA	NA	NA
Egbin	Egbin	GFST	5,385,475.96	1320	980.89	0.74	6	1986	32.8
Geregu	Kogi	SCGT	776,765.16	414	300	0.72	3	2007	8.2
Ibom	Ikot Abasi	SCGT	264,931.66	190	2.90	0.08	NA	NA	NA
Okpai	Okpai	CCGT	3,232,402	480	446.25	0.93	NA	NA	NA
Olorunsogo	Olorunsogo	SCGT	270,295.90	336	315.75	0.94	NA	2008	2.3
Omoku	Omoku	SCGT	217,145.11	150	96.05	0.96	NA	NA	NA
Omotosho	Omotosho	SCGT	144,212	336	298.15	0.89	NA	2007	2.3
Sapele	Sapele	GFST	513,636.84	1020	199.07	0.20	10	1978	2.7
Trans Amadi	Rivers	SCGT	75,220.16	136	NA	NA	NA	NA	NA
Ughelli	Delta	SCGT	1,957,869.10	900	255.33	0.29	20	1966	8.9
Total			17,360,102.31	6902	3517				59
Hydro									
Kainji	Niger	Hydro	2,300,991	800	445.45	0.59	12	1968	14.3
Jebba	Niger	Hydro	2,693,741	540	457.99	0.79	6	1985	13.3
Shiroro	Kaduna	Hydro	2,421,116	600	404.56	0.67	6	1989	12.4
Total			7,415,848	1940	1308				41

Source Power Holding Company of Nigeria PLC (PHCN) (2010) National Control Centre Osogbo Generation and Transmission Grid Operations 2010 Annual Technical Report. Osogbo, Nigeria (Unpublished Data)

Table B.28 Categories of energy saving strategies (2010)

Mode	Electricity (%)	Share of total energy consumed in BTUs						Total (%)
		Diesel fuel (%)	Gasoline (%)	LNG and blends (%)	CNG and blends (%)	Biodiesel (%)	Others (%)	
Bus	0	68	1	4	21	6	1	100
Commuter rail	33	66	0	0	0	0	1	100
Heavy rail	100	0	0	0	0	0	0	100
Light rail	95	5	0	0	0	0	0	100
Paratransit	0	39	545	0	2	4	1	100
Trolley bus	100	0	0	0	0	0	0	100
Other	3	69	28	0	0	0	0	100
All modes	15	56	10	2	12	4	1	100

Source Gallivan (2013)

Table B.29 Technology cost for vehicle fuels (Switching/Conversion) (2010)

Fuel type	Cost (US$)	Number of years to annualize cost to vehicles (years/Vel km)
Petrol	0.07	10
Diesel	0.05	10
CNG	0.03	15
LPG	0.03	16

Source Rahman (2009)

Appendix C
Sectorial Energy Demand by Scenarios

Table C.1 Reference scenario energy demand by sectors

Energy demand final units							
Reference scenario, all fuels							
Branch: Demand							
Units: Petajoules							
Branches	2010	2015	2020	2025	2030	2035	2040
Households	418.5	502.4	603.1	724.2	869.6	1,049.8	1,267.6
Urban	207.2	254.6	312.6	383.6	470.6	580.1	714.7
Electrified	173.9	218.9	275.3	345.7	433.5	546.0	686.9
Lighting	33.8	42.3	52.9	66.0	82.3	102.6	127.6
Electric	11.7	14.6	18.1	22.5	27.8	34.4	42.5
LPG	5.0	6.3	7.9	9.9	12.4	15.5	19.4
Kerosene	16.4	20.6	25.8	32.3	40.5	50.6	63.1
Other	0.7	0.8	1.0	1.3	1.6	2.0	2.5
Cooking	93.2	117.5	147.8	185.7	233.0	293.3	368.8
Electric	3.2	4.0	5.0	6.3	7.8	9.2	10.8
Kerosene	53.8	67.6	84.8	106.3	133.0	166.2	207.4
LPG	36.1	45.8	57.8	73.0	92.1	117.7	150.3
Firewood	0.1	0.1	0.1	0.1	0.1	0.2	0.2
Air-conditioning	2.7	3.5	4.6	6.0	7.7	10.4	13.8
Existing	2.7	3.5	4.6	6.0	7.7	10.4	13.8
Refrigeration	11.3	14.5	18.6	23.7	30.3	39.3	50.8
Existing	11.3	14.5	18.6	23.7	30.3	39.3	50.8
Food preservation	23.4	29.2	36.2	44.9	55.6	68.8	85.1

(continued)

N.V. Emodi, *Energy Policies for Sustainable Development Strategies*,
Frontiers in African Business Research, DOI 10.1007/978-981-10-0974-7

Table C.1 (continued)

Energy demand final units							
Reference scenario, all fuels							
Branch: Demand							
Units: Petajoules							
Branches	2010	2015	2020	2025	2030	2035	2040
Existing	23.4	29.2	36.2	44.9	55.6	68.8	85.1
Water heating	0.1	0.2	0.2	0.2	0.3	0.4	0.4
Existing	0.1	0.2	0.2	0.2	0.3	0.4	0.4
Electronics	9.3	11.8	15.0	19.1	24.2	31.3	40.3
Existing	9.3	11.8	15.0	19.1	24.2	31.3	40.3
Non-electrified	33.3	35.6	37.3	38.0	37.1	34.1	27.8
Lighting	8.9	9.5	9.8	10.0	9.7	8.8	7.1
Electric	2.9	3.1	3.2	3.2	3.1	2.8	2.2
LPG	0.2	0.2	0.3	0.3	0.2	0.2	0.2
Kerosene	5.7	6.0	6.3	6.4	6.2	5.7	4.6
Other	0.1	0.1	0.1	0.1	0.1	0.1	0.1
Cooking	19.4	20.7	21.6	22.0	21.4	19.6	15.9
Electric	0.4	0.4	0.4	0.4	0.4	0.3	0.2
Kerosene	12.8	13.7	14.3	14.5	14.1	12.8	10.4
LPG	6.1	6.5	6.9	7.0	6.9	6.4	5.3
Firewood	0.1	0.1	0.1	0.1	0.1	0.1	0.0
Air-conditioning	0.2	0.3	0.3	0.3	0.4	0.4	0.4
Existing	0.2	0.3	0.3	0.3	0.4	0.4	0.4
Refrigeration	0.9	1.1	1.2	1.3	1.3	1.3	1.1
Existing	0.9	1.1	1.2	1.3	1.3	1.3	1.1
Food preservation	2.9	3.1	3.2	3.2	3.1	2.8	2.2
Existing	2.9	3.1	3.2	3.2	3.1	2.8	2.2
Water heating	0.0	0.0	0.0	0.0	0.0	0.0	0.0
Existing	0.0	0.0	0.0	0.0	0.0	0.0	0.0
Electronics	0.9	1.0	1.1	1.2	1.2	1.2	1.1
Existing	0.9	1.0	1.1	1.2	1.2	1.2	1.1
Rural	211.4	247.8	290.5	340.6	399.1	469.7	552.8
Electrified	73.6	104.9	144.9	195.5	259.1	340.6	442.7
Lighting	36.5	51.8	71.2	95.7	126.3	164.4	211.6
Electric	3.9	5.5	7.5	10.0	13.1	16.8	21.5
Kerosene	32.5	46.2	63.5	85.4	112.8	147.0	189.5
Other	0.1	0.2	0.2	0.3	0.4	0.5	0.6
Cooking	32.5	46.2	63.6	85.5	112.9	147.1	189.5
Electric	0.1	0.1	0.1	0.1	0.2	0.2	0.2
Kerosene	29.8	42.3	58.2	78.2	103.3	134.6	173.5

(continued)

Table C.1 (continued)

Energy demand final units							
Reference scenario, all fuels							
Branch: Demand							
Units: Petajoules							
Branches	2010	2015	2020	2025	2030	2035	2040
LPG	0.4	0.5	0.7	0.9	1.2	1.6	2.1
Firewood	2.0	2.9	4.0	5.3	7.1	9.2	11.8
Charcoal	0.3	0.4	0.6	0.8	1.0	1.3	1.7
Sawdust_Plant Residue	0.0	0.0	0.0	0.1	0.1	0.1	0.1
Air-conditioning	0.3	0.6	0.9	1.4	2.1	3.4	5.2
Existing	0.3	0.6	0.9	1.4	2.1	3.4	5.2
Refrigeration	0.5	0.9	1.5	2.4	3.6	6.2	9.8
Existing	0.5	0.9	1.5	2.4	3.6	6.2	9.8
Food preservation	2.6	3.7	5.1	6.9	9.1	11.8	15.2
Existing	2.6	3.7	5.1	6.9	9.1	11.8	15.2
Water heating	0.0	0.0	0.0	0.0	0.0	0.0	0.1
Existing	0.0	0.0	0.0	0.0	0.0	0.0	0.1
Electronics	1.0	1.6	2.5	3.6	5.2	7.8	11.3
Existing	1.0	1.6	2.5	3.6	5.2	7.8	11.3
Non-electrified	137.8	142.9	145.7	145.1	140.0	129.1	110.2
Lighting	61.4	63.5	64.5	64.0	61.6	56.5	47.9
Electric	4.8	5.0	5.0	4.9	4.7	4.3	3.6
Kerosene	56.3	58.3	59.3	58.9	56.7	52.1	44.2
Other	0.2	0.2	0.2	0.2	0.2	0.2	0.1
Cooking	75.0	77.7	79.0	78.5	75.5	69.3	58.9
Kerosene	60.5	62.6	63.7	63.3	60.9	55.9	47.5
Firewood	13.2	13.7	13.9	13.8	13.3	12.2	10.4
Charcoal	1.2	1.3	1.3	1.3	1.2	1.1	1.0
Sawdust plant residue	0.1	0.1	0.1	0.1	0.1	0.1	0.1
Refrigeration	0.4	0.6	0.7	0.9	1.0	1.2	1.2
Existing	0.4	0.6	0.7	0.9	1.0	1.2	1.2
Electronics	1.0	1.2	1.5	1.7	1.9	2.1	2.1
Existing	1.0	1.2	1.5	1.7	1.9	2.1	2.1
Industry	297.0	346.0	403.0	469.5	546.9	637.1	742.2
Industry	297.0	346.0	403.0	469.5	546.9	637.1	742.2
Residual fuel oil	11.0	13.6	16.7	20.5	25.0	30.5	37.1
Natural gas	25.4	33.3	43.2	55.3	70.3	88.7	111.3

(continued)

Table C.1 (continued)

Energy demand final units							
Reference scenario, all fuels							
Branch: Demand							
Units: Petajoules							
Branches	2010	2015	2020	2025	2030	2035	2040
Biomass waste	247.9	273.8	301.5	330.9	361.8	393.9	426.8
Electricity	11.7	22.9	37.5	56.2	80.1	110.4	148.4
Coal	1.0	2.4	4.2	6.6	9.7	13.6	18.6
Commercial service	84.0	112.4	150.4	201.3	269.4	360.5	482.5
Commercial service	84.0	112.4	150.4	201.3	269.4	360.5	482.5
Residual fuel oil	0.3	0.3	0.5	0.6	0.8	1.1	1.4
Biomass waste	64.1	85.8	114.9	153.7	205.7	275.3	368.4
Electricity	19.6	26.2	35.1	47.0	62.9	84.2	112.7
Agriculture	0.4	0.5	0.7	0.9	1.1	1.4	1.8
Agriculture	0.4	0.5	0.7	0.9	1.1	1.4	1.8
Kerosene	0.1	0.1	0.1	0.2	0.2	0.3	0.4
Diesel	0.3	0.4	0.5	0.7	0.9	1.1	1.4
Transport	239.3	277.5	321.7	372.9	432.3	501.1	581.0
Road vehicles	239.3	277.5	321.7	372.9	432.3	501.1	581.0
Motorcycles	2.4	2.8	3.2	3.7	4.3	5.0	5.8
Gasoline moto	2.4	2.8	3.2	3.7	4.3	5.0	5.8
Cars	229.6	266.1	308.5	357.7	414.6	480.7	557.2
Gasoline cars	187.7	217.6	252.3	292.4	339.0	393.0	455.6
Diesel cars	41.9	48.5	56.2	65.2	75.6	87.6	101.6
Light goods vehicles	2.0	2.3	2.7	3.1	3.6	4.2	4.9
Gasoline LGV	2.0	2.3	2.7	3.1	3.6	4.2	4.9
Heavy goods vehicles	0.3	0.4	0.5	0.5	0.6	0.7	0.8
Diesel HGV	0.3	0.4	0.5	0.5	0.6	0.7	0.8
Urban bus	1.7	2.0	2.3	2.6	3.1	3.6	4.1
Gasoline UB	1.7	2.0	2.3	2.6	3.1	3.6	4.1
Long-distance coach	3.3	3.9	4.5	5.2	6.0	7.0	8.1
Diesel_LDC	3.3	3.9	4.5	5.2	6.0	7.0	8.1
Total	1,039.3	1,238.8	1,478.9	1,768.8	2,119.4	2,550.0	3,075.0

Table C.2 Low carbon moderate scenario energy demand by sectors

Energy demand final units							
Low carbon moderate scenario, all fuels							
Branch: Demand							
Units: Petajoules							
Branches	2010	2015	2020	2025	2030	2035	2040
Households	418.5	496.4	588.2	696.4	823.8	977.3	1,157.4
Urban	207.2	248.5	297.6	355.9	424.7	507.6	604.6
Electrified	173.9	212.9	260.3	317.9	387.7	484.9	604.6
Lighting	33.8	40.3	47.9	56.8	67.2	81.4	98.0
Electric	11.7	13.5	15.5	17.7	19.9	22.8	25.7
LPG	5.0	6.2	7.5	9.2	11.2	14.1	17.6
Kerosene	16.4	19.8	23.8	28.7	34.5	42.5	52.2
Other	0.7	0.8	1.0	1.2	1.5	2.0	2.5
Cooking	93.2	113.9	138.8	169.1	205.7	256.7	319.3
Electric	3.2	4.0	4.9	6.0	7.3	9.1	11.4
Kerosene	53.8	65.0	78.3	94.2	113.2	139.6	171.6
LPG	36.1	44.9	55.6	68.8	85.0	107.8	136.1
Firewood	0.1	0.1	0.1	0.1	0.1	0.2	0.2
Air-conditioning	2.7	3.5	4.6	5.9	7.6	10.1	13.3
Efficient	2.7	3.5	4.6	5.9	7.6	10.1	13.3
Refrigeration	11.3	14.0	17.3	21.4	26.3	33.1	41.5
Efficient	11.3	14.0	17.3	21.4	26.3	33.1	41.5
Food preservation	23.4	29.2	36.2	44.9	55.6	70.7	89.6
Efficient	23.4	29.2	36.2	44.9	55.6	70.7	89.6
Water heating	0.1	0.2	0.2	0.2	0.3	0.4	0.4
Efficient	0.1	0.2	0.2	0.2	0.3	0.4	0.4
Electronics	9.3	11.9	15.3	19.5	24.9	32.6	42.5
Efficient	9.3	11.9	15.3	19.5	24.9	32.6	42.5
Non-electrified	33.3	35.6	37.3	38.0	37.1	22.7	–
Lighting	8.9	9.5	9.8	10.0	9.7	5.8	–
Electric	2.9	3.1	3.2	3.2	3.1	1.9	–
LPG	0.2	0.2	0.3	0.3	0.2	0.2	–
Kerosene	5.7	6.0	6.3	6.4	6.2	3.8	–
Other	0.1	0.1	0.1	0.1	0.1	0.1	–
Cooking	19.4	20.7	21.6	22.0	21.4	13.1	–
Electric	0.4	0.4	0.4	0.4	0.4	0.2	–
Kerosene	12.8	13.7	14.3	14.5	14.1	8.6	–
LPG	6.1	6.5	6.9	7.0	6.9	4.3	–
Firewood	0.1	0.1	0.1	0.1	0.1	0.0	–

(continued)

Table C.2 (continued)

Energy demand final units							
Low carbon moderate scenario, all fuels							
Branch: Demand							
Units: Petajoules							
Branches	2010	2015	2020	2025	2030	2035	2040
Air-conditioning	0.2	0.3	0.3	0.3	0.4	0.3	–
Efficient	0.2	0.3	0.3	0.3	0.4	0.3	–
Refrigeration	0.9	1.1	1.2	1.3	1.3	0.9	–
Efficient	0.9	1.1	1.2	1.3	1.3	0.9	–
Food preservation	2.9	3.1	3.2	3.2	3.1	1.9	–
Efficient	2.9	3.1	3.2	3.2	3.1	1.9	–
Water heating	0.0	0.0	0.0	0.0	0.0	0.0	–
Efficient	0.0	0.0	0.0	0.0	0.0	0.0	–
Electronics	0.9	1.0	1.1	1.2	1.2	0.8	–
Efficient	0.9	1.0	1.1	1.2	1.2	0.8	–
Rural	211.4	247.8	290.5	340.6	399.1	469.7	552.8
Electrified	73.6	104.9	144.9	195.5	259.1	340.6	442.7
Lighting	36.5	51.8	71.2	95.7	126.3	164.4	211.6
Electric	3.9	5.5	7.5	10.0	13.1	16.8	21.5
Kerosene	32.5	46.2	63.5	85.4	112.8	147.0	189.5
Other	0.1	0.2	0.2	0.3	0.4	0.5	0.6
Cooking	32.5	46.2	63.6	85.5	112.9	147.1	189.5
Electric	0.1	0.1	0.1	0.1	0.2	0.2	0.2
Kerosene	29.8	42.3	58.2	78.2	103.3	134.6	173.5
LPG	0.4	0.5	0.7	0.9	1.2	1.6	2.1
Firewood	2.0	2.9	4.0	5.3	7.1	9.2	11.8
Charcoal	0.3	0.4	0.6	0.8	1.0	1.3	1.7
Sawdust plant residue	0.0	0.0	0.0	0.1	0.1	0.1	0.1
Air-conditioning	0.3	0.6	0.9	1.4	2.1	3.4	5.2
Efficient	0.3	0.6	0.9	1.4	2.1	3.4	5.2
Refrigeration	0.5	0.9	1.5	2.4	3.6	6.2	9.8
Efficient	0.5	0.9	1.5	2.4	3.6	6.2	9.8
Food preservation	2.6	3.7	5.1	6.9	9.1	11.8	15.2
Efficient	2.6	3.7	5.1	6.9	9.1	11.8	15.2
Water heating	0.0	0.0	0.0	0.0	0.0	0.0	0.1
Efficient	0.0	0.0	0.0	0.0	0.0	0.0	0.1
Electronics	1.0	1.6	2.5	3.6	5.2	7.8	11.3

(continued)

Table C.2 (continued)

Energy demand final units							
Low carbon moderate scenario, all fuels							
Branch: Demand							
Units: Petajoules							
Branches	2010	2015	2020	2025	2030	2035	2040
Efficient	1.0	1.6	2.5	3.6	5.2	7.8	11.3
Non-electrified	137.8	142.9	145.7	145.1	140.0	129.1	110.2
Lighting	61.4	63.5	64.5	64.0	61.6	56.5	47.9
Electric	4.8	5.0	5.0	4.9	4.7	4.3	3.6
Kerosene	56.3	58.3	59.3	58.9	56.7	52.1	44.2
Other	0.2	0.2	0.2	0.2	0.2	0.2	0.1
Cooking	75.0	77.7	79.0	78.5	75.5	69.3	58.9
Kerosene	60.5	62.6	63.7	63.3	60.9	55.9	47.5
Firewood	13.2	13.7	13.9	13.8	13.3	12.2	10.4
Charcoal	1.2	1.3	1.3	1.3	1.2	1.1	1.0
Sawdust plant residue	0.1	0.1	0.1	0.1	0.1	0.1	0.1
Refrigeration	0.4	0.6	0.7	0.9	1.0	1.2	1.2
Efficient	0.4	0.6	0.7	0.9	1.0	1.2	1.2
Electronics	1.0	1.2	1.5	1.7	1.9	2.1	2.1
Efficient	1.0	1.2	1.5	1.7	1.9	2.1	2.1
Industry	297.0	337.4	383.3	435.5	494.8	562.1	638.6
Industry	297.0	337.4	383.3	435.5	494.8	562.1	638.6
Residual fuel oil	11.0	13.3	15.9	19.0	22.6	26.9	31.9
Natural gas	25.4	32.5	41.0	51.3	63.6	78.3	95.8
Biomass waste	247.9	267.0	286.8	306.9	327.3	347.5	367.2
Electricity	11.7	22.3	35.6	52.1	72.5	97.4	127.7
Coal	1.0	2.3	4.0	6.1	8.8	12.0	16.0
Commercial service	84.0	111.3	147.4	195.4	258.8	342.9	454.3
Commercial service	84.0	111.3	147.4	195.4	258.8	342.9	454.3
Residual fuel oil	0.3	0.3	0.4	0.6	0.8	1.0	1.4
Biomass waste	64.1	82.8	106.9	137.8	177.5	228.5	294.0
Electricity	19.6	28.1	40.2	57.0	80.5	113.4	159.0
Agriculture	0.4	0.5	0.7	0.9	1.1	1.4	1.8
Agriculture	0.4	0.5	0.7	0.9	1.1	1.4	1.8
Kerosene	0.1	0.1	0.1	0.2	0.2	0.3	0.4
Diesel	0.3	0.4	0.5	0.7	0.9	1.1	1.4

(continued)

Table C.2 (continued)

Energy demand final units							
Low carbon moderate scenario, all fuels							
Branch: Demand							
Units: Petajoules							
Branches	2010	2015	2020	2025	2030	2035	2040
Transport	239.3	285.7	340.9	406.6	484.8	577.8	688.3
Road vehicles	239.3	285.7	340.9	406.6	484.8	577.8	688.3
Motorcycles	2.4	2.8	3.4	4.0	4.8	5.7	6.7
Gasoline moto	2.4	2.8	3.3	3.9	4.5	5.3	6.2
Bio moto	–	0.0	0.1	0.1	0.2	0.3	0.5
Cars	229.6	273.9	326.8	389.7	464.4	553.3	658.9
Gasoline cars	187.7	212.7	239.7	268.5	298.4	328.7	358.1
Diesel cars	41.9	48.1	55.1	62.8	71.3	80.6	90.5
Bio cars	–	13.1	32.0	58.4	94.7	144.0	210.3
Light goods vehicles	2.0	2.4	2.9	3.5	4.3	5.2	6.2
Gasoline LGV	2.0	2.3	2.7	3.1	3.5	4.0	4.6
Bio LGV	–	0.1	0.3	0.5	0.8	1.2	1.7
Heavy goods vehicles	0.3	0.4	0.5	0.6	0.7	0.9	1.0
Diesel HGV	0.3	0.4	0.4	0.5	0.6	0.6	0.7
Bio HGV	–	0.0	0.1	0.1	0.2	0.2	0.4
Urban bus	1.7	2.0	2.5	3.0	3.6	4.3	5.2
Gasoline UB	1.7	1.9	2.1	2.3	2.5	2.6	2.8
Bio UB	–	0.2	0.4	0.7	1.1	1.7	2.5
Long-distance coach	3.3	4.0	4.8	5.8	7.0	8.5	10.2
Diesel LDC	3.3	3.8	4.3	4.8	5.4	5.9	6.5
Bio LDC	–	0.2	0.6	1.0	1.7	2.5	3.7
Total	1,039.3	1,231.3	1,460.6	1,734.8	2,063.3	2,461.6	2,940.5

Table C.3 Low carbon advance scenario energy demand by sectors

Energy demand final units							
Low carbon advance scenario, all fuels							
Branch: Demand							
Units: Petajoules							
Branches	2010	2015	2020	2025	2030	2035	2040
Households	418.5	506.4	608.6	723.7	848.6	925.1	948.3
Urban	207.2	261.6	323.6	390.5	458.1	464.2	403.6
Electrified	173.9	189.9	211.8	240.3	276.8	333.0	403.6
Lighting	33.8	39.2	45.3	52.1	59.7	70.0	81.4
Electric	11.7	12.8	13.8	14.6	15.2	15.7	15.6
LPG	5.0	6.2	7.6	9.4	11.5	14.6	18.3
Kerosene	16.4	19.4	22.9	27.0	31.7	38.2	45.8
Other	0.7	0.8	0.9	1.1	1.3	1.5	1.8
Cooking	93.2	96.9	104.0	115.1	131.1	158.2	194.4
Electric	3.2	4.0	5.0	6.3	7.8	9.5	11.4
Kerosene	53.8	47.0	41.1	35.8	31.1	27.8	24.8
LPG	36.1	45.8	57.8	73.0	92.1	120.9	158.2
Firewood	0.1	0.0	0.0	0.0	0.0	0.0	0.0
Air-conditioning	2.7	2.7	2.7	2.7	2.7	2.9	3.0
Efficient	2.7	2.7	2.7	2.7	2.7	2.9	3.0
Refrigeration	11.3	10.8	10.4	9.9	9.4	9.3	9.2
Efficient	11.3	10.8	10.4	9.9	9.4	9.3	9.2
Food preservation	23.4	28.7	35.1	42.9	52.4	65.6	81.8
Efficient	23.4	28.7	35.1	42.9	52.4	65.6	81.8
Water heating	0.1	0.2	0.2	0.2	0.3	0.3	0.4
Efficient	0.1	0.2	0.2	0.2	0.3	0.3	0.4
Electronics	9.3	11.4	14.1	17.3	21.2	26.6	33.3
Efficient	9.3	11.4	14.1	17.3	21.2	26.6	33.3
Non-electrified	33.3	71.7	111.8	150.2	181.3	131.2	–
Lighting	8.9	45.5	84.4	122.2	153.9	114.4	–
Electric	2.9	39.2	77.7	115.5	147.3	110.4	–
LPG	0.2	0.2	0.3	0.3	0.2	0.2	–
Kerosene	5.7	6.0	6.3	6.4	6.2	3.8	–
Other	0.1	0.1	0.1	0.1	0.1	0.1	–
Cooking	19.4	20.7	21.6	22.0	21.4	13.1	–
Electric	0.4	0.4	0.4	0.4	0.4	0.2	–
Kerosene	12.8	13.7	14.3	14.5	14.1	8.6	–
LPG	6.1	6.5	6.9	7.0	6.9	4.3	–
Firewood	0.1	0.1	0.1	0.1	0.1	0.0	–

(continued)

Table C.3 (continued)

Energy demand final units							
Low carbon advance scenario, all fuels							
Branch: Demand							
Units: Petajoules							
Branches	2010	2015	2020	2025	2030	2035	2040
Air-conditioning	0.2	0.3	0.3	0.3	0.4	0.3	–
Efficient	0.2	0.3	0.3	0.3	0.4	0.3	–
Refrigeration	0.9	1.1	1.2	1.3	1.3	0.9	–
Efficient	0.9	1.1	1.2	1.3	1.3	0.9	–
Food preservation	2.9	3.1	3.2	3.2	3.1	1.9	–
Efficient	2.9	3.1	3.2	3.2	3.1	1.9	–
Water heating	0.0	0.0	0.0	0.0	0.0	0.0	–
Efficient	0.0	0.0	0.0	0.0	0.0	0.0	–
Electronics	0.9	1.0	1.1	1.2	1.2	0.8	–
Efficient	0.9	1.0	1.1	1.2	1.2	0.8	–
Rural	211.4	244.7	285.0	333.1	390.6	460.9	544.7
Electrified	73.6	104.9	144.9	195.5	259.1	340.6	442.7
Lighting	36.5	51.8	71.2	95.7	126.3	164.4	211.6
Electric	3.9	5.5	7.5	10.0	13.1	16.8	21.5
Kerosene	32.5	46.2	63.5	85.4	112.8	147.0	189.5
Other	0.1	0.2	0.2	0.3	0.4	0.5	0.6
Cooking	32.5	46.2	63.6	85.5	112.9	147.1	189.5
Electric	0.1	0.1	0.1	0.1	0.2	0.2	0.2
Kerosene	29.8	42.3	58.2	78.2	103.3	134.6	173.5
LPG	0.4	0.5	0.7	0.9	1.2	1.6	2.1
Firewood	2.0	2.9	4.0	5.3	7.1	9.2	11.8
Charcoal	0.3	0.4	0.6	0.8	1.0	1.3	1.7
Sawdust Plant Residue	0.0	0.0	0.0	0.1	0.1	0.1	0.1
Air-conditioning	0.3	0.6	0.9	1.4	2.1	3.4	5.2
Efficient	0.3	0.6	0.9	1.4	2.1	3.4	5.2
Refrigeration	0.5	0.9	1.5	2.4	3.6	6.2	9.8
Efficient	0.5	0.9	1.5	2.4	3.6	6.2	9.8
Food preservation	2.6	3.7	5.1	6.9	9.1	11.8	15.2
Efficient	2.6	3.7	5.1	6.9	9.1	11.8	15.2
Water heating	0.0	0.0	0.0	0.0	0.0	0.0	0.1
Efficient	0.0	0.0	0.0	0.0	0.0	0.0	0.1
Electronics	1.0	1.6	2.5	3.6	5.2	7.8	11.3

(continued)

Table C.3 (continued)

Energy demand final units							
Low carbon advance scenario, all fuels							
Branch: Demand							
Units: Petajoules							
Branches	2010	2015	2020	2025	2030	2035	2040
Efficient	1.0	1.6	2.5	3.6	5.2	7.8	11.3
Non-electrified	137.8	139.8	140.1	137.7	131.4	120.3	102.0
Lighting	61.4	63.5	64.5	64.0	61.6	56.5	47.9
Electric	4.8	5.0	5.0	4.9	4.7	4.3	3.6
Kerosene	56.3	58.3	59.3	58.9	56.7	52.1	44.2
Other	0.2	0.2	0.2	0.2	0.2	0.2	0.1
Cooking	75.0	74.6	73.4	71.1	67.0	60.5	50.7
Kerosene	60.5	62.6	63.7	63.3	60.9	55.9	47.5
Firewood	13.2	10.6	8.3	6.4	4.8	3.4	2.2
Charcoal	1.2	1.3	1.3	1.3	1.2	1.1	1.0
Sawdust plant residue	0.1	0.1	0.1	0.1	0.1	0.1	0.1
Refrigeration	0.4	0.6	0.7	0.9	1.0	1.2	1.2
Efficient	0.4	0.6	0.7	0.9	1.0	1.2	1.2
Electronics	1.0	1.2	1.5	1.7	1.9	2.1	2.1
Efficient	1.0	1.2	1.5	1.7	1.9	2.1	2.1
Industry	297.0	320.8	346.5	374.3	404.3	436.6	471.6
Industry	297.0	320.8	346.5	374.3	404.3	436.6	471.6
Residual fuel oil	11.0	12.6	14.4	16.3	18.5	20.9	23.6
Natural gas	25.4	36.3	48.6	62.8	78.9	97.2	117.9
Biomass waste	247.9	238.1	225.1	208.6	187.9	162.6	132.1
Electricity	11.7	31.9	55.3	82.2	113.1	148.4	188.7
Coal	1.0	1.9	3.0	4.3	5.8	7.5	9.4
Commercial service	84.0	111.3	147.4	195.4	258.8	342.9	454.3
Commercial service	84.0	111.3	147.4	195.4	258.8	342.9	454.3
Residual fuel oil	0.3	0.4	0.6	0.9	1.3	1.9	2.7
Biomass waste	64.1	80.4	100.5	125.1	155.1	191.4	234.9
Electricity	19.6	30.5	46.4	69.4	102.5	149.7	216.7
Agriculture	0.4	0.5	0.7	0.9	1.1	1.4	1.8
Agriculture	0.4	0.5	0.7	0.9	1.1	1.4	1.8
Kerosene	0.1	0.1	0.1	0.2	0.2	0.3	0.4
Diesel	0.3	0.4	0.5	0.7	0.9	1.1	1.4
Transport	239.3	280.9	329.3	385.5	450.5	525.7	612.2

(continued)

Table C.3 (continued)

Energy demand final units							
Low carbon advance scenario, all fuels							
Branch: Demand							
Units: Petajoules							
Branches	2010	2015	2020	2025	2030	2035	2040
Road vehicles	239.3	280.9	329.3	385.5	450.5	525.7	612.2
Motorcycles	2.4	2.8	3.3	3.9	4.5	5.3	6.2
Gasoline moto	2.4	2.8	3.2	3.7	4.2	4.8	5.4
Bio moto	–	0.0	0.1	0.2	0.3	0.5	0.7
Cars	229.6	269.4	315.7	369.5	431.7	503.5	586.2
Gasoline cars	187.7	205.1	221.2	234.7	243.6	245.3	236.4
Diesel cars	41.9	46.4	50.9	55.3	59.1	62.0	63.3
Bio cars	–	9.9	24.0	43.8	71.0	108.0	157.7
LPG cars	–	8.1	19.6	35.8	58.0	88.2	128.8
Light goods vehicles	2.0	2.4	2.8	3.3	3.9	4.6	5.4
Gasoline LGV	2.0	2.2	2.3	2.3	2.3	2.2	2.0
Bio LGV	–	0.1	0.3	0.5	0.8	1.2	1.7
LPG LGV	–	0.1	0.3	0.5	0.8	1.2	1.7
Heavy goods vehicles	0.3	0.4	0.5	0.6	0.7	0.9	1.0
Diesel HGV	0.3	0.4	0.4	0.5	0.5	0.5	0.6
Bio HGV	–	0.0	0.1	0.1	0.2	0.3	0.5
Urban bus	1.7	2.0	2.4	2.8	3.4	4.0	4.7
Gasoline UB	1.7	1.9	2.0	2.1	2.2	2.3	2.2
Bio UB	–	0.1	0.2	0.4	0.7	1.0	1.5
LPG UB	–	0.1	0.2	0.3	0.4	0.7	1.0
Long-distance coach	3.3	3.9	4.6	5.4	6.4	7.5	8.7
Diesel LDC	3.3	3.6	3.9	4.1	4.1	4.1	3.8
Bio LDC	–	0.2	0.4	0.8	1.2	1.9	2.8
LPG LDC	–	0.1	0.3	0.6	1.0	1.5	2.2
Total	1,039.3	1,219.9	1,432.5	1,679.7	1,963.4	2,231.7	2,488.3

Table C.4 Green optimistic scenario energy demand by sectors

Energy demand final units							
GO scenario, all fuels							
Branch: Demand							
Units: Petajoules							
Branches	2010	2015	2020	2025	2030	2035	2040
Households	418.5	459.6	510.1	570.9	643.4	732.4	836.0
Urban	207.2	211.8	219.5	230.4	244.3	262.6	283.1
Electrified	173.9	181.6	191.9	205.3	221.9	249.8	283.1
Lighting	33.8	36.4	39.1	41.9	44.7	48.9	53.0
Electric	11.7	11.4	11.0	10.4	9.7	9.0	8.0
LPG	5.0	6.6	8.6	11.1	14.2	18.6	24.2
Kerosene	16.4	17.6	18.7	19.5	19.9	20.2	19.7
Other	0.7	0.7	0.8	0.9	1.0	1.1	1.1
Cooking	93.2	92.4	93.2	95.6	99.7	109.2	120.9
Electric	3.2	3.9	4.8	5.8	7.1	8.4	9.8
Kerosene	53.8	47.0	41.1	35.8	31.1	27.8	24.8
LPG	36.1	41.4	47.3	53.9	61.5	73.0	86.3
Firewood	0.1	0.0	0.0	0.0	0.0	0.0	0.0
Air-conditioning	2.7	2.6	2.4	2.3	2.2	2.2	2.2
Efficient	2.7	2.6	2.4	2.3	2.2	2.2	2.2
Refrigeration	11.3	11.1	10.6	9.9	8.9	8.0	6.8
Efficient	11.3	11.1	10.6	9.9	8.9	8.0	6.8
Food preservation	23.4	28.7	35.1	42.9	52.4	65.6	81.8
Efficient	23.4	28.7	35.1	42.9	52.4	65.6	81.8
Water heating	0.1	0.2	0.2	0.2	0.3	0.3	0.4
Efficient	0.1	0.2	0.2	0.2	0.3	0.3	0.4
Electronics	9.3	10.2	11.3	12.4	13.7	15.8	18.1
Efficient	9.3	10.2	11.3	12.4	13.7	15.8	18.1
Non-electrified	33.3	30.2	27.6	25.1	22.4	12.8	–
Lighting	8.9	9.5	9.8	10.0	9.7	5.8	–
Electric	2.9	3.1	3.2	3.2	3.1	1.9	–
LPG	0.2	0.2	0.3	0.3	0.2	0.2	–
Kerosene	5.7	6.0	6.3	6.4	6.2	3.8	–
Other	0.1	0.1	0.1	0.1	0.1	0.1	–
Cooking	19.4	15.3	11.9	9.1	6.8	3.2	–
Electric	0.4	0.4	0.4	0.4	0.4	0.2	–
Kerosene	12.8	9.5	6.9	4.9	3.3	1.4	–
LPG	6.1	5.3	4.6	3.8	3.1	1.5	–
Firewood	0.1	0.0	0.0	0.0	0.0	0.0	–
Air-conditioning	0.2	0.3	0.3	0.3	0.4	0.3	–
Efficient	0.2	0.3	0.3	0.3	0.4	0.3	–

(continued)

Table C.4 (continued)

Energy demand final units							
GO scenario, all fuels							
Branch: Demand							
Units: Petajoules							
Branches	2010	2015	2020	2025	2030	2035	2040
Refrigeration	0.9	1.1	1.2	1.3	1.3	0.9	–
Efficient	0.9	1.1	1.2	1.3	1.3	0.9	–
Food preservation	2.9	3.1	3.2	3.2	3.1	1.9	–
Efficient	2.9	3.1	3.2	3.2	3.1	1.9	–
Water heating	0.0	0.0	0.0	0.0	0.0	0.0	–
Efficient	0.0	0.0	0.0	0.0	0.0	0.0	–
Electronics	0.9	1.0	1.1	1.2	1.2	0.8	–
Efficient	0.9	1.0	1.1	1.2	1.2	0.8	–
Rural	211.4	247.8	290.5	340.6	399.1	469.7	552.8
Electrified	73.6	104.9	144.9	195.5	259.1	340.6	442.7
Lighting	36.5	51.8	71.2	95.7	126.3	164.4	211.6
Electric	3.9	5.5	7.5	10.0	13.1	16.8	21.5
Kerosene	32.5	46.2	63.5	85.4	112.8	147.0	189.5
Other	0.1	0.2	0.2	0.3	0.4	0.5	0.6
Cooking	32.5	46.2	63.6	85.5	112.9	147.1	189.5
Electric	0.1	0.1	0.1	0.1	0.2	0.2	0.2
Kerosene	29.8	42.3	58.2	78.2	103.3	134.6	173.5
LPG	0.4	0.5	0.7	0.9	1.2	1.6	2.1
Firewood	2.0	2.9	4.0	5.3	7.1	9.2	11.8
Charcoal	0.3	0.4	0.6	0.8	1.0	1.3	1.7
Sawdust plant residue	0.0	0.0	0.0	0.1	0.1	0.1	0.1
Air-conditioning	0.3	0.6	0.9	1.4	2.1	3.4	5.2
Efficient	0.3	0.6	0.9	1.4	2.1	3.4	5.2
Refrigeration	0.5	0.9	1.5	2.4	3.6	6.2	9.8
Efficient	0.5	0.9	1.5	2.4	3.6	6.2	9.8
Food preservation	2.6	3.7	5.1	6.9	9.1	11.8	15.2
Efficient	2.6	3.7	5.1	6.9	9.1	11.8	15.2
Water heating	0.0	0.0	0.0	0.0	0.0	0.0	0.1
Efficient	0.0	0.0	0.0	0.0	0.0	0.0	0.1
Electronics	1.0	1.6	2.5	3.6	5.2	7.8	11.3
Efficient	1.0	1.6	2.5	3.6	5.2	7.8	11.3
Non-electrified	137.8	142.9	145.7	145.1	140.0	129.1	110.2
Lighting	61.4	63.5	64.5	64.0	61.6	56.5	47.9
Electric	4.8	5.0	5.0	4.9	4.7	4.3	3.6
Kerosene	56.3	58.3	59.3	58.9	56.7	52.1	44.2

(continued)

Table C.4 (continued)

Energy demand final units							
GO scenario, all fuels							
Branch: Demand							
Units: Petajoules							
Branches	2010	2015	2020	2025	2030	2035	2040
Other	0.2	0.2	0.2	0.2	0.2	0.2	0.1
Cooking	75.0	77.7	79.0	78.5	75.5	69.3	58.9
Kerosene	60.5	62.6	63.7	63.3	60.9	55.9	47.5
Firewood	13.2	13.7	13.9	13.8	13.3	12.2	10.4
Charcoal	1.2	1.3	1.3	1.3	1.2	1.1	1.0
Sawdust plant residue	0.1	0.1	0.1	0.1	0.1	0.1	0.1
Refrigeration	0.4	0.6	0.7	0.9	1.0	1.2	1.2
Efficient	0.4	0.6	0.7	0.9	1.0	1.2	1.2
Electronics	1.0	1.2	1.5	1.7	1.9	2.1	2.1
Efficient	1.0	1.2	1.5	1.7	1.9	2.1	2.1
Industry	297.0	312.7	329.3	346.8	365.1	384.5	404.9
Industry	297.0	312.7	329.3	346.8	365.1	384.5	404.9
Residual fuel oil	11.0	10.7	10.4	9.9	9.4	8.8	8.1
Natural gas	25.4	37.9	51.7	66.9	83.4	101.6	121.5
Biomass waste	247.9	225.8	200.8	172.4	140.5	104.7	64.8
Electricity	11.7	36.3	63.5	93.5	126.5	162.7	202.4
Coal	1.0	1.9	2.9	4.0	5.3	6.6	8.1
Commercial service	84.0	111.3	147.4	195.4	258.8	342.9	454.3
Commercial service	84.0	111.3	147.4	195.4	258.8	342.9	454.3
Residual fuel oil	0.3	0.5	0.9	1.6	2.5	3.9	5.9
Biomass waste	64.1	77.3	92.3	108.8	126.3	143.7	159.0
Electricity	19.6	33.5	54.3	85.0	130.1	195.4	289.4
Agriculture	0.4	0.5	0.7	0.9	1.1	1.4	1.8
Agriculture	0.4	0.5	0.7	0.9	1.1	1.4	1.8
Kerosene	0.1	0.1	0.1	0.2	0.2	0.3	0.4
Diesel	0.3	0.4	0.5	0.7	0.9	1.1	1.4
Transport	239.3	277.2	320.2	368.8	423.5	484.6	552.2
Road vehicles	239.3	277.2	320.2	368.8	423.5	484.6	552.2
Motorcycles	2.4	2.8	3.2	3.7	4.3	4.9	5.6
Gasoline moto	2.4	2.7	3.1	3.4	3.8	4.3	4.7
Bio moto	–	0.1	0.2	0.3	0.4	0.7	1.0
Cars	229.6	265.8	306.9	353.4	405.6	463.9	528.4
Gasoline cars	187.7	199.7	208.1	210.8	204.8	186.4	150.4
Diesel cars	41.9	44.7	46.8	47.7	46.9	43.4	36.2
Bio cars	–	9.9	24.0	43.8	71.0	108.0	157.7
LPG cars	–	6.7	16.3	29.8	48.3	73.5	107.3

(continued)

Table C.4 (continued)

Energy demand final units							
GO scenario, all fuels							
Branch: Demand							
Units: Petajoules							
Branches	2010	2015	2020	2025	2030	2035	2040
CNG cars	–	4.8	11.7	21.3	34.5	52.5	76.7
Light goods vehicles	2.0	2.4	2.8	3.2	3.8	4.4	5.1
Gasoline LGV	2.0	2.2	2.3	2.3	2.3	2.2	2.0
Bio LGV	–	0.1	0.2	0.3	0.5	0.8	1.1
LPG LGV	–	0.1	0.2	0.3	0.5	0.7	1.1
CNG LGV	–	0.1	0.1	0.3	0.4	0.7	1.0
Heavy goods vehicles	0.3	0.4	0.5	0.6	0.7	0.8	0.9
Diesel HGV	0.3	0.4	0.4	0.4	0.4	0.4	0.4
Bio HGV	–	0.0	0.0	0.1	0.1	0.2	0.3
CNG HGV	–	0.0	0.0	0.1	0.1	0.2	0.2
Urban Bus	1.7	2.0	2.3	2.7	3.1	3.6	4.2
Gasoline UB	1.7	1.8	1.9	2.0	2.0	1.9	1.7
Bio UB	–	0.1	0.2	0.3	0.4	0.7	1.0
LPG UB	–	0.1	0.1	0.2	0.4	0.6	0.8
CNG UB	–	0.0	0.1	0.2	0.3	0.5	0.7
Long-distance coach	3.3	3.9	4.5	5.2	6.0	6.9	8.0
Diesel LDC	3.3	3.6	3.8	3.9	3.9	3.7	3.3
Bio LDC	–	0.1	0.3	0.5	0.8	1.3	1.8
LPG LDC	–	0.1	0.2	0.4	0.7	1.1	1.5
CNG LDC	–	0.1	0.2	0.4	0.6	0.9	1.3
Total	1,039.3	1,161.4	1,307.7	1,482.7	1,692.0	1,945.8	2,249.2

Appendix D
Primary and Secondary Energy Supply by Scenarios

Table D.1 Total primary energy requirements by scenarios

Primary requirements							
All fuels							
Branch: Resources							
Units: Million Gigajoules							
Scenarios	2010	2015	2020	2025	2030	2035	2040
GO	1,270.6	1,568.0	1,801.2	2,139.0	2,504.8	2,968.5	3,507.5
Low carbon advance	1,270.6	1,601.3	1,934.1	2,337.8	2,795.7	3,185.7	3,486.7
Low carbon moderate	1,270.6	1,584.5	1,872.4	2,221.4	2,669.0	3,202.5	3,868.9
Reference	1,270.6	1,631.2	1,930.6	2,309.7	2,755.9	3,326.3	3,998.0
Total	5,082.4	6,385.0	7,538.4	9,007.8	10,725.5	12,683.0	14,861.0

N.V. Emodi, *Energy Policies for Sustainable Development Strategies*,
Frontiers in African Business Research, DOI 10.1007/978-981-10-0974-7

Table D.2 Reference scenario fuel supply

Energy demand final units							
Reference scenario							
Branch: Demand							
Units: Petajoules							
Fuels	2010	2015	2020	2025	2030	2035	2040
Electricity	116.2	155.1	204.6	267.5	347.0	450.9	582.5
Natural gas	25.4	33.3	43.2	55.3	70.3	88.7	111.3
Gasoline	193.8	224.7	260.5	301.9	350.0	405.8	470.4
Kerosene	267.9	317.4	376.0	445.5	527.8	625.2	740.6
Diesel	45.9	53.2	61.7	71.6	83.1	96.5	112.0
Residual fuel oil	11.3	13.9	17.2	21.1	25.8	31.6	38.6
LPG	47.8	59.4	73.6	91.2	112.9	141.5	177.2
Coal sub-bituminous	1.0	2.4	4.2	6.6	9.7	13.6	18.6
Wood	15.4	16.7	18.0	19.3	20.6	21.6	22.5
Charcoal	1.5	1.7	1.9	2.1	2.3	2.5	2.7
Vegetal wastes	0.1	0.1	0.1	0.1	0.2	0.2	0.2
Solar	1.0	1.2	1.3	1.5	1.6	1.8	1.9
Biomass	312.0	359.6	416.4	484.6	567.5	669.2	795.1
Total	1,039.3	1,238.8	1,478.9	1,768.8	2,119.4	2,550.0	3,075.0

Table D.3 Low carbon moderate scenario fuel supply

Energy demand final units							
Low carbon moderate scenario							
Branch: Demand							
Units: Petajoules							
Fuels	2010	2015	2020	2025	2030	2035	2040
Electricity	116.2	155.0	204.1	266.3	345.2	449.2	581.3
Natural gas	25.4	32.5	41.0	51.3	63.6	78.3	95.8
Gasoline	193.8	219.7	247.8	277.7	308.9	340.7	371.6
Kerosene	267.9	313.9	367.5	429.8	502.0	584.4	678.8
Diesel	45.9	52.7	60.3	68.8	78.1	88.3	99.1
Residual fuel oil	11.3	13.6	16.3	19.6	23.4	27.9	33.3
LPG	47.8	58.3	71.0	86.2	104.7	127.9	155.8
Coal Sub-bituminous	1.0	2.3	4.0	6.1	8.8	12.0	16.0
Wood	15.4	16.7	18.0	19.3	20.5	21.6	22.4
Charcoal	1.5	1.7	1.9	2.1	2.3	2.5	2.7
Vegetal wastes	0.1	0.1	0.1	0.1	0.2	0.2	0.2
Solar	1.0	1.2	1.4	1.6	1.9	2.2	2.6
Biomass	312.0	363.5	426.9	505.5	603.5	726.1	880.1
Total	1,039.3	1,231.3	1,460.6	1,734.8	2,063.3	2,461.6	2,940.5

Table D.4 Low carbon advance scenario fuel supply

Energy demand final units							
Low carbon advance scenario							
Branch: Demand							
Units: Petajoules							
Fuels	2010	2015	2020	2025	2030	2035	2040
Electricity	116.2	197.4	291.9	399.4	518.9	596.1	630.4
Natural gas	25.4	36.3	48.6	62.8	78.9	97.2	117.9
Gasoline	193.8	211.9	228.7	242.8	252.3	254.6	246.0
Kerosene	267.9	295.7	329.4	369.7	417.2	468.3	525.5
Diesel	45.9	50.8	55.8	60.5	64.6	67.7	69.2
Residual fuel oil	11.3	13.0	15.0	17.2	19.8	22.8	26.3
LPG	47.8	67.6	93.7	127.8	172.2	233.1	312.2
Coal sub-bituminous	1.0	1.9	3.0	4.3	5.8	7.5	9.4
Wood	15.4	13.6	12.4	11.8	11.9	12.6	14.1
Charcoal	1.5	1.7	1.9	2.1	2.3	2.5	2.7
Vegetal wastes	0.1	0.1	0.1	0.1	0.2	0.2	0.2
Solar	1.0	1.2	1.5	1.8	2.2	2.7	3.3
Biomass	312.0	328.8	350.7	379.4	417.3	466.9	531.8
Total	1,039.3	1,219.9	1,432.5	1,679.7	1,963.4	2,231.7	2,488.3

Table D.5 Green optimistic scenario fuel supply

Energy demand final units							
Green optimistic advance scenario							
Branch: Demand							
Units: Petajoules							
Fuels	2010	2015	2020	2025	2030	2035	2040
Electricity	116.2	166.1	227.6	304.1	400.9	526.9	689.1
Natural gas	25.4	37.9	51.7	66.9	83.4	101.6	121.5
Gasoline	193.8	206.4	215.4	218.6	213.0	194.8	158.7
Kerosene	267.9	289.7	317.8	352.5	394.5	443.1	499.4
Diesel	45.9	49.1	51.5	52.8	52.1	48.7	41.3
Residual fuel oil	11.3	11.3	11.3	11.5	11.9	12.7	14.0
LPG	47.8	61.0	78.2	100.8	130.2	170.8	223.3
Coal sub bituminous	1.0	1.9	2.9	4.0	5.3	6.6	8.1
Wood	15.4	16.7	18.0	19.2	20.4	21.4	22.2
Charcoal	1.5	1.7	1.9	2.1	2.3	2.5	2.7
Vegetal wastes	0.1	0.1	0.1	0.1	0.2	0.2	0.2
Solar	1.0	1.3	1.5	1.9	2.3	2.8	3.4
Biomass	312.0	313.3	317.8	326.4	340.2	360.0	386.7
CNG	–	5.0	12.2	22.2	36.0	54.8	80.0
Total	1,039.3	1,161.4	1,307.7	1,482.7	1,692.0	1,945.8	2,249.2

Appendix E
Total Electricity Supply by Scenarios

Table E.1 Reference scenario electricity supply

Outputs by feedstock fuel							
Reference scenario, all fuels, primary outputs							
Branch: Transformation\Electricity Generation\Processes							
Units: Million Megawatt-Hours							
Branches	2010	2015	2020	2025	2030	2035	2040
Hydropower_1	7.4	3.1	6.0	7.4	7.0	7.2	9.3
Coal_Steam	–	0.7	1.2	4.0	5.0	9.5	13.4
Nuclear	–	0.8	1.3	1.9	2.8	3.8	4.6
On_Shore Wind	–	–	0.0	0.0	0.0	0.0	0.0
Biomass	–	0.0	0.0	0.0	0.0	0.0	0.1
Solar PV	–	0.0	0.0	0.1	0.1	0.1	0.2
Solar Thermal	–	0.0	0.0	0.0	0.0	0.0	0.0
SCGT_1	11.2	29.7	33.7	38.6	45.3	55.8	64.2
CCGT_1	6.2	3.5	5.0	6.8	17.9	23.8	32.2
Diesel Gen_Off Grid	–	8.3	9.5	13.5	16.0	22.2	30.4
Gasoline Gen_Off Grid	–	4.7	5.6	7.0	6.8	7.0	8.9
Gas Turbine_Off Grid	–	1.9	5.9	8.0	10.1	12.4	16.5
Total	24.8	52.8	68.2	87.4	111.2	141.8	179.8

N.V. Emodi, *Energy Policies for Sustainable Development Strategies*,
Frontiers in African Business Research, DOI 10.1007/978-981-10-0974-7

Table E.2 Low carbon moderate scenario electricity supply

Outputs by feedstock fuel							
Low carbon moderate scenario, all fuels, primary outputs							
Branch: Transformation\Electricity Generation\Processes							
Units: Million Megawatt-hours							
Branches	2010	2015	2020	2025	2030	2035	2040
Hydropower_1	7.4	3.9	4.6	6.0	7.7	8.8	9.7
Coal_Steam	–	1.3	2.3	3.3	6.6	8.3	13.6
Nuclear	–	0.7	1.3	2.4	3.7	4.6	5.5
On_Shore Wind	–	0.0	0.0	0.0	0.0	0.0	0.1
Biomass	–	0.3	0.5	0.9	1.1	1.6	2.0
Solar PV	–	0.0	0.0	0.6	1.2	2.0	2.7
Solar thermal	–	0.0	0.1	0.1	0.3	0.4	0.5
Coal_Super_Critical	–	0.7	1.2	1.7	4.7	7.7	11.4
Solar PV_Off Grid	–	0.0	0.1	0.2	0.5	0.9	1.3
Small Hydropower_Off Grid	–	1.1	2.0	2.3	2.9	3.3	3.5
SCGT_1	11.2	17.4	18.2	19.6	21.0	27.9	38.3
CCGT_1	6.2	13.8	17.8	19.4	22.2	23.3	28.9
Diesel Gen_Off Grid	–	7.3	8.5	12.0	13.5	19.0	20.2
Gasoline Gen_Off Grid	–	4.1	6.4	8.0	11.5	13.9	16.7
Gas Turbine_Off Grid	–	1.7	4.0	8.5	10.4	14.5	17.5
Total	24.8	52.3	67.0	85.0	107.3	136.1	171.8

Table E.3 Low carbon advance scenario electricity supply

outputs by feedstock fuel							
Low carbon advance scenario, all fuels, primary outputs							
Branch: Transformation\Electricity Generation\Processes							
Units: Million Megawatt-Hours							
Branches	2010	2015	2020	2025	2030	2035	2040
Hydropower_1	7.4	16.3	17.6	16.6	16.7	15.5	15.4
Nuclear	–	1.3	2.5	6.2	9.4	14.3	18.6
On_Shore Wind	–	0.0	0.1	1.5	2.7	3.8	4.7
Biomass	–	0.6	1.1	3.8	6.9	7.3	8.6
Solar PV	–	0.2	0.3	1.3	1.8	3.1	4.2
Solar thermal	–	0.2	0.3	0.8	2.2	4.1	5.2
Coal_IGCC	–	1.2	2.3	7.8	12.4	13.5	15.2
Coal_CFB	–	2.5	7.0	9.8	12.4	15.0	16.4
Solar PV_Off Grid	–	0.2	0.4	0.5	1.7	2.2	2.1
Small Hydropower_Off Grid	–	2.5	4.7	6.9	8.0	9.0	8.8
On_Shore Wind_Off Grid	–	0.3	0.6	1.8	2.7	3.4	3.9
SCGT_1	11.2	18.3	19.8	18.7	18.8	17.5	14.9
CCGT_1	6.2	16.6	23.5	30.6	37.7	41.9	39.3
Gas Turbine_Off Grid	–	2.3	6.6	9.2	11.7	12.7	12.1
SCGT_Off Grid	–	3.9	8.6	11.4	15.0	15.9	14.9
Total	24.8	66.5	95.4	126.8	160.2	179.0	184.3

Table E.4 Green optimistic scenario electricity supply

Outputs by feedstock fuel							
GO scenario, all fuels, primary outputs							
Branch: Transformation\Electricity Generation\Processes							
Units: Million Megawatt-hours							
Branches	2010	2015	2020	2025	2030	2035	2040
Hydropower_1	7.4	6.4	11.2	11.6	12.8	16.3	20.1
Nuclear	–	2.4	4.2	9.8	14.4	19.8	26.2
On_Shore Wind	–	0.0	0.5	3.1	4.8	6.8	8.3
Biomass	–	0.9	1.6	3.8	6.0	8.5	10.7
Solar PV	–	0.9	1.5	2.4	3.2	5.2	6.1
Solar thermal	–	1.1	2.0	3.8	4.7	5.9	7.5
Coal_IGCC	–	1.1	2.0	3.1	5.4	8.0	11.0
Coal_Super_Critical	–	1.1	2.0	3.1	4.1	5.3	6.9
Coal_CFB	–	2.3	5.0	6.2	8.2	9.3	13.1
Off_Shore Wind	–	0.1	0.1	1.6	3.6	4.9	6.6
Geothermal	–	1.2	2.1	4.1	5.8	8.5	12.4
Solar PV_Off Grid	–	0.3	0.5	1.2	1.8	2.4	3.2
Small Hydropower_Off Grid	–	2.3	4.0	5.4	7.5	9.1	11.0
On_Shore Wind_Off Grid	–	0.3	0.5	1.6	2.8	4.1	5.6
Biomass_Off Grid	–	0.6	1.0	3.0	4.7	5.9	8.1
SCGT_1	11.2	16.1	14.7	11.9	11.3	11.3	12.4
CCGT_1	6.2	15.2	17.7	17.2	18.3	20.9	24.6
SCGT_Off Grid	–	3.6	3.6	3.3	3.6	4.7	5.8
Total	24.8	55.8	74.1	96.0	122.8	156.8	199.4

Appendix F
Total Electricity Demand, Supply and Primary Resources Requirements by Scenarios

Table F.1 Total electricity demand

Energy demand final units							
Electricity							
Branch: Demand							
Units: Million Megawatt-hours							
Scenarios	2010	2015	2020	2025	2030	2035	2040
GO	32.3	46.2	63.2	84.5	111.4	146.3	191.4
Low carbon advance	32.3	54.8	81.1	111.0	144.1	165.6	175.1
Low carbon moderate	32.3	43.0	56.7	74.0	95.9	124.8	161.5
Reference	32.3	43.1	56.8	74.3	96.4	125.2	161.8
Total	129.1	187.1	257.8	343.7	447.8	561.9	689.8

Table F.2 Total electricity supply

Outputs by feedstock fuel							
All fuels, primary outputs							
Branch: Transformation\Electricity Generation							
Units: Million Megawatt-hours							
Scenarios	2010	2015	2020	2025	2030	2035	2040
GO	24.8	55.8	74.1	96.0	122.8	156.8	199.4
Low carbon advance	24.8	66.5	95.4	126.8	160.2	179.0	184.3
Low carbon moderate	24.8	52.3	67.0	85.0	107.3	136.1	171.8
Reference	24.8	52.8	68.2	87.4	111.2	141.8	179.8
Total	99.1	227.3	304.6	395.3	501.5	613.7	735.3

N.V. Emodi, *Energy Policies for Sustainable Development Strategies*,
Frontiers in African Business Research, DOI 10.1007/978-981-10-0974-7

Table F.3 Primary resource requirements

Primary requirements							
All fuels							
Branch: Resources							
Units: Petajoules							
Scenarios	2010	2015	2020	2025	2030	2035	2040
GO	1,270.6	1,568.0	1,801.2	2,139.0	2,504.8	2,968.5	3,507.5
Low carbon advance	1,270.6	1,601.3	1,934.1	2,337.8	2,795.7	3,185.7	3,486.7
Low carbon moderate	1,270.6	1,584.5	1,872.4	2,221.4	2,669.0	3,202.5	3,868.9
Reference	1,270.6	1,631.2	1,930.6	2,309.7	2,755.9	3,326.3	3,998.0
Total	5,082.4	6,385.0	7,538.4	9,007.8	10,725.5	12,683.0	14,861.0

Appendix G
Green House Gas (GHG) Emissions by Scenarios

Table G.1 GHGs in reference scenario

One hundred year global warming potential							
Reference scenario, all fuels, all GHGs							
Branch: Demand							
Units: Million metric tons CO_2 equivalent							
Branches	2010	2015	2020	2025	2030	2035	2040
Households	22.3	26.6	31.7	37.7	45.0	53.7	64.1
Urban	9.1	11.2	13.6	16.7	20.4	25.0	30.6
Rural	13.2	15.4	18.0	21.1	24.6	28.7	33.5
Industry	2.8	3.6	4.6	5.8	7.3	9.2	11.5
Industry	2.8	3.6	4.6	5.8	7.3	9.2	11.5
Commercial service	0.5	0.7	0.9	1.2	1.6	2.2	2.9
Commercial service	0.5	0.7	0.9	1.2	1.6	2.2	2.9
Agriculture	0.0	0.0	0.0	0.1	0.1	0.1	0.1
Agriculture	0.0	0.0	0.0	0.1	0.1	0.1	0.1
Transport	16.8	19.4	22.5	26.1	30.3	35.1	40.7
Road vehicles	16.8	19.4	22.5	26.1	30.3	35.1	40.7
Total	42.4	50.3	59.7	70.9	84.2	100.2	119.2

N.V. Emodi, *Energy Policies for Sustainable Development Strategies*,
Frontiers in African Business Research, DOI 10.1007/978-981-10-0974-7

Table G.2 GHGs in low carbon moderate scenario

One hundred year global warming potential							
Low carbon moderate scenario, all fuels, all GHGs							
Branch: Demand							
Units: Million metric tons CO_2 equivalent							
Branches	2010	2015	2020	2025	2030	2035	2040
Households	22.3	26.3	30.9	36.3	42.6	49.9	58.4
Urban	9.1	10.9	12.9	15.3	18.0	21.2	24.9
Rural	13.2	15.4	18.0	21.1	24.6	28.7	33.5
Industry	2.8	3.5	4.4	5.4	6.6	8.1	9.9
Industry	2.8	3.5	4.4	5.4	6.6	8.1	9.9
Commercial service	0.5	0.6	0.8	1.1	1.4	1.8	2.3
Commercial service	0.5	0.6	0.8	1.1	1.4	1.8	2.3
Agriculture	0.0	0.0	0.0	0.1	0.1	0.1	0.1
Agriculture	0.0	0.0	0.0	0.1	0.1	0.1	0.1
Transport	16.8	19.1	21.8	24.7	27.8	31.1	34.5
Road vehicles	16.8	19.1	21.8	24.7	27.8	31.1	34.5
Total	42.4	49.6	58.0	67.5	78.5	91.0	105.2

Table G.3 GHGs in low advance moderate scenario

One hundred year global warming potential							
Low carbon advance scenario, all fuels, all GHGs							
Branch: Demand							
Units: Million metric tons CO_2 equivalent							
Branches	2010	2015	2020	2025	2030	2035	2040
Households	22.3	25.0	28.2	32.1	36.8	42.2	48.4
Urban	9.1	9.6	10.2	11.1	12.3	13.5	15.0
Rural	13.2	15.4	18.0	21.0	24.5	28.6	33.4
Industry	2.8	3.6	4.5	5.5	6.7	8.0	9.4
Industry	2.8	3.6	4.5	5.5	6.7	8.0	9.4
Commercial service	0.5	0.6	0.8	1.0	1.3	1.6	2.0
Commercial service	0.5	0.6	0.8	1.0	1.3	1.6	2.0
Agriculture	0.0	0.0	0.0	0.1	0.1	0.1	0.1
Agriculture	0.0	0.0	0.0	0.1	0.1	0.1	0.1
Transport	16.8	18.9	21.2	23.6	26.1	28.6	30.8
Road vehicles	16.8	18.9	21.2	23.6	26.1	28.6	30.8
Total	42.4	48.1	54.8	62.3	70.9	80.4	90.8

Table G.4 GHGs in green optimistic scenario

One hundred year global warming potential							
Green optimistic scenario, all fuels, all GHGs							
Branch: Demand							
Units: Million metric tons CO_2 equivalent							
Branches	2010	2015	2020	2025	2030	2035	2040
Households	22.3	24.3	26.7	29.8	33.4	37.8	42.9
Urban	9.1	8.8	8.7	8.7	8.8	9.1	9.4
Rural	13.2	15.4	18.0	21.1	24.6	28.7	33.5
Industry	2.8	3.5	4.3	5.2	6.1	7.1	8.3
Industry	2.8	3.5	4.3	5.2	6.1	7.1	8.3
Commercial service	0.5	0.6	0.8	0.9	1.1	1.4	1.6
Commercial service	0.5	0.6	0.8	0.9	1.1	1.4	1.6
Agriculture	0.0	0.0	0.0	0.1	0.1	0.1	0.1
Agriculture	0.0	0.0	0.0	0.1	0.1	0.1	0.1
Transport	16.8	18.6	20.5	22.3	23.9	25.2	26.0
Road vehicles	16.8	18.6	20.5	22.3	23.9	25.2	26.0
Total	42.4	47.1	52.3	58.2	64.7	71.6	78.9

Table G.5 GHG from electricity generation in reference scenario

One hundred year global warming potential							
Reference scenario, all fuels, all GHGs							
Branch: Transformation\Electricity Generation\Processes							
Units: Million metric tons CO_2 equivalent							
Branches	2010	2015	2020	2025	2030	2035	2040
Coal_Steam	–	0.8	1.3	4.5	5.6	10.7	15.0
Biomass	–	0.0	0.0	0.0	0.0	0.0	0.0
SCGT_1	5.6	14.9	16.9	19.4	22.8	28.0	32.3
CCGT_1	2.1	1.2	1.7	2.4	6.2	8.2	11.2
Diesel Gen_Off Grid	–	3.1	3.6	5.0	6.0	8.3	11.3
Gasoline Gen_Off Grid	–	3.5	4.2	5.3	5.1	5.2	6.6
Gas Turbine_Off Grid	–	0.6	2.0	2.7	3.4	4.2	5.5
Total	7.8	24.2	29.7	39.2	49.1	64.6	81.9

Table G.6 GHG from electricity generation in low carbon moderate scenario

One hundred year global warming potential							
Reference scenario, all fuels, all GHGs							
Branch: Transformation\Electricity Generation\Processes							
Units: Million metric tons CO_2 equivalent							
Branches	2010	2015	2020	2025	2030	2035	2040
Coal_Steam	–	1.4	2.6	3.7	7.4	9.2	15.2
Biomass	–	0.0	0.0	0.0	0.0	0.0	0.0
Coal_Super_Critical	–	0.6	1.1	1.5	4.0	6.6	9.8
SCGT_1	5.6	8.7	9.1	9.8	10.6	14.0	19.2
CCGT_1	2.1	4.8	6.2	6.7	7.7	8.1	10.0
Diesel Gen_Off Grid	–	2.7	3.2	4.5	5.0	7.1	7.6
Gasoline Gen_Off Grid	–	3.1	4.8	6.0	8.6	10.4	12.5
Gas Turbine_Off Grid	–	0.6	1.4	2.8	3.5	4.8	5.9
Total	7.8	21.9	28.2	35.0	46.8	60.3	80.1

Table G.7 GHG from electricity generation in low carbon advance scenario

One hundred year global warming potential							
Reference scenario, all fuels, all GHGs							
Branch: Transformation\Electricity Generation\Processes							
Units: Million metric tons CO_2 equivalent							
Branches	2010	2015	2020	2025	2030	2035	2040
Biomass	–	0.0	0.0	0.0	0.1	0.1	0.1
Coal_IGCC	–	0.9	1.7	5.8	9.2	10.0	11.3
Coal_CFB	–	2.6	7.3	10.3	13.0	15.7	17.2
SCGT_1	5.6	9.2	9.9	9.4	9.4	8.8	7.5
CCGT_1	2.1	5.8	8.1	10.6	13.1	14.5	13.6
Gas Turbine_Off Grid	–	0.8	2.2	3.1	3.9	4.3	4.1
SCGT_Off Grid	–	2.0	4.3	5.7	7.6	8.0	7.5
Total	7.8	21.2	33.7	44.9	56.3	61.3	61.2

Table G.8 GHG from electricity generation in green optimistic scenario

One hundred year global warming potential							
Reference scenario, all fuels, all GHGs							
Branch: Transformation\Electricity Generation\Processes							
Units: Million metric tons CO_2 equivalent							
Branches	2010	2015	2020	2025	2030	2035	2040
Biomass	–	0.0	0.0	0.0	0.1	0.1	0.1
Coal_IGCC	–	0.8	1.5	2.3	4.1	6.0	8.2
Coal_Super_Critical	–	1.0	1.7	2.6	3.5	4.6	5.9
Coal_CFB	–	2.4	5.2	6.5	8.6	9.8	13.7
SCGT_1	5.6	8.1	7.4	6.0	5.7	5.7	6.2
CCGT_1	2.1	5.3	6.1	6.0	6.3	7.2	8.5
SCGT_Off Grid		1.8	1.8	1.6	1.8	2.3	2.9
Total	7.8	19.4	23.7	25.0	30.0	35.7	45.6

Appendix H
Energy Balance by Scenarios

N.V. Emodi, *Energy Policies for Sustainable Development Strategies*,
Frontiers in African Business Research, DOI 10.1007/978-981-10-0974-7

Table H.1 Reference scenario energy balance in base year (2010)

Energy balance for area "Nigeria"

Scenario: Reference, year: 2010, Units: Petajoule	Electricity	Natural gas	Gasoline	Kerosene	Diesel	Residual fuel oil	LPG	Crude oil	Coal sub bituminous	Wood	Charcoal	Vegetal wastes	Solar	Hydro	Lubricants	Biomass	Avgas	CNG	Total
Production	–	186.1	–	–	–	–	–	963.9	0.8	23.0	–	0.1	1.0	28.1	–	312.0	–	–	1,515.0
Imports	56.0	–	85.4	159.5	–	–	–	–	0.3	–	–	–	–	–	–	–	–	–	301.2
Exports	–	–	–	–	−62.6	−97.1	−60.6	–	–	–	–	–	–	–	−108.4	–	−108.4	−108.4	−545.6
Total primary supply	56.0	186.1	85.4	159.5	−62.6	−97.1	−60.6	963.9	1.1	23.0	–	0.1	1.0	28.1	−108.4	312.0	−108.4	−108.4	1,270.6
Coal mining	–	–	–	–	–	–	–	–	−0.2	–	–	–	–	–	–	–	–	–	−0.2
Oil refining	–	–	108.4	108.4	108.4	108.4	108.4	−963.9	–	–	–	–	–	–	108.4	–	108.4	108.4	−96.4
Charcoal production	–	–	–	–	–	–	–	–	–	−7.6	1.5	–	–	–	–	–	–	–	−6.1
Electricity generation	89.2	−139.0	–	–	–	–	–	–	–	–	–	–	–	−28.1	–	–	–	–	−77.9
Transmission and distribution	−29.0	−21.7	–	–	–	–	–	–	–	–	–	–	–	–	–	–	–	–	−50.7
Total transformation	60.1	−160.7	108.4	108.4	108.4	108.4	108.4	−963.9	−0.2	−7.6	1.5	–	–	−28.1	108.4	–	108.4	108.4	−231.3
Households	84.9	–	–	267.8	–	–	47.8	–	–	15.4	1.5	0.1	1.0	–	–	–	–	–	418.5
Industry	11.7	25.4	–	–	–	11.0	–	–	1.0	–	–	–	–	–	–	247.9	–	–	297.0
Commercial_ Service	19.6	–	–	–	–	0.3	–	–	–	–	–	–	–	–	–	64.1	–	–	84.0
Agriculture	–	–	–	0.1	0.3	–	–	–	–	–	–	–	–	–	–	–	–	–	0.4
Transport	–	–	193.8	–	45.5	–	–	–	–	–	–	–	–	–	–	–	–	–	239.3
Total demand	116.2	25.4	193.8	267.9	45.9	11.3	47.8	–	1.0	15.4	1.5	0.1	1.0	–	–	312.0	–	–	1,039.3
Unmet requirements (Waste)	–	–	–	–	–	–	–	–	–	–	–	–	–	–	–	–	–	–	–

Table H.2 Reference scenario energy balance in 2040

Energy Balance for Area "Nigeria"

Scenario: Reference, Year: 2040, Units: Petajoule	Electricity	Natural gas	Gasoline	Kerosene	Diesel	Residual fuel oil	LPG	Crude oil	Coal sub bituminous	Wood	Charcoal	Vegetal wastes	Wind	Solar	Hydro	Nuclear	Lubricants	Biomass	Avgas	CNG	Total
Production	–	1,016.1	–	–	–	–	–	963.9	0.8	34.6	–	0.2	0.2	7.8	35.2	–	–	795.4	–	–	2,854.1
Imports	–	–	453.0	632.1	159.7	–	68.8	–	179.0	–	–	–	–	–	–	46.5	–	–	–	–	1,539.1
Exports	–	–	–	–	–	−69.9	–	–	–	–	–	–	–	–	–	–	−108.4	–	−108.4	−108.4	−395.2
Total primary supply	–	1,016.1	453.0	632.1	159.7	−69.9	68.8	963.9	179.8	34.6	–	0.2	0.2	7.8	35.2	46.5	−108.4	795.4	−108.4	−108.4	3,998.0
Coal mining	–	–	–	–	–	–	–	–	−0.2	–	–	–	–	–	–	–	–	–	–	–	−0.2
Oil refining	–	–	108.4	108.4	108.4	108.4	108.4	−963.9	–	–	–	–	–	–	–	–	108.4	–	108.4	108.4	−96.4
Charcoal production	–	–	–	–	–	–	–	–	–	−12.1	2.7	–	–	–	–	–	–	–	–	–	−9.4
Electricity generation	647.2	−876.9	−91.1	–	−156.1	–	–	–	−161.1	–	–	–	−0.2	−4.4	−35.2	−46.5	–	−0.3	–	–	−724.5
Transmission and distribution	−64.7	−27.8	–	–	–	–	–	–	–	–	–	–	–	–	–	–	–	–	–	–	−92.6
Total Transformation	582.5	−904.7	17.4	108.4	−47.7	108.4	108.4	−963.9	−161.2	−12.1	2.7	–	−0.2	−4.4	−35.2	−46.5	108.4	−0.3	108.4	108.4	−923.0
Households	321.4	–	–	740.2	–	–	177.2	–	–	22.5	2.7	0.2	–	3.4	–	–	–	–	–	–	1,267.6
Industry	148.4	111.3	–	–	–	37.1	–	–	18.6	–	–	–	–	–	–	–	–	426.8	–	–	742.2
Commercial Service	112.7	–	–	–	–	1.4	–	–	–	–	–	–	–	–	–	–	–	368.4	–	–	482.5
Agriculture	–	–	–	0.4	1.4	–	–	–	–	–	–	–	–	–	–	–	–	–	–	–	1.8
Transport	–	–	470.4	–	110.5	–	–	–	–	–	–	–	–	–	–	–	–	–	–	–	581.0
Total demand	582.5	111.3	470.4	740.6	112.0	38.6	177.2	–	18.6	22.5	2.7	0.2	–	3.4	–	–	–	795.1	–	–	3,075.0
Unmet require-ments (Waste)	–	–	–	–	–	–	–	–	–	–	–	–	–	–	–	–	–	–	–	–	–

Table H.3 Low carbon moderate scenario energy balance in 2040

Energy balance for area "Nigeria"																					
Scenario: Low Carbon Moderate, Year: 2040, Units: Petajoule	Electricity	Natural gas	Gasoline	Kerosene	Diesel	Residual fuel oil	LPG	Crude oil	Coal sub bituminous	Wood	Charcoal	Vegetal wastes	Wind	Solar	Hydro	Nuclear	Lubricants	Biomass	Avgas	CNG	Total
Production	–	738.6	–	–	–	–	–	963.9	0.8	32.2	–	0.2	0.7	107.5	50.9	–	–	889.2	–	–	2,784.0
Imports	–	–	434.7	570.4	94.8	–	47.3	–	283.5	–	–	–	–	–	–	54.7	–	–	–	–	1,485.3
Exports	–	–	–	–	–	−75.1	–	–	–	–	–	–	–	–	–	–	−108.4	–	−108.4	−108.4	−400.5
Total primary supply	–	738.6	434.7	570.4	94.8	−75.1	47.3	963.9	284.3	32.2	–	0.2	0.7	107.5	50.9	54.7	−108.4	889.2	−108.4	−108.4	3,868.9
Coal mining	–	–	–	–	–	–	–	–	−0.2	–	–	–	–	–	–	–	–	–	–	–	−0.2
Oil refining	–	–	108.4	108.4	108.4	108.4	108.4	−963.9	–	–	–	–	–	–	–	–	108.4	–	108.4	108.4	−96.4
Charcoal production	–	–	–	–	–	–	–	–	–	−9.8	2.7	–	–	–	–	–	–	–	–	–	−7.1
Electricity generation	618.4	−628.5	−171.5	–	−104.1	–	–	–	−268.2	–	–	–	−0.7	−104.2	−50.9	−54.7	–	−9.1	–	–	−773.3
Transmission and distribution	−37.1	−14.3	–	–	–	–	–	–	–	–	–	–	–	–	–	–	–	–	–	–	−51.4
Total Transformation	581.3	−642.8	−63.0	108.4	4.3	108.4	108.4	−963.9	−268.3	−9.8	2.7	–	−0.7	−104.2	−50.9	−54.7	108.4	−9.1	108.4	108.4	−928.4
Households	294.6	–	–	678.5	–	–	155.8	–	–	22.4	2.7	0.2	–	3.3	–	–	–	–	–	–	1,157.4
Industry	127.7	95.8	–	–	–	31.9	–	–	16.0	–	–	–	–	–	–	–	–	367.2	–	–	638.6
Commercial Service	159.0	–	–	–	–	1.4	–	–	–	–	–	–	–	–	–	–	–	294.0	–	–	454.3
Agriculture	–	–	–	0.4	1.4	–	–	–	–	–	–	–	–	–	–	–	–	–	–	–	1.8
Transport	–	–	371.6	–	97.7	–	–	–	–	–	–	–	–	–	–	–	–	219.0	–	–	688.3
Total demand	581.3	95.8	371.6	678.8	99.1	33.3	155.8	–	16.0	22.4	2.7	0.2	–	3.3	–	–	–	880.1	–	–	2,940.5
Unmet requirements (Waste)	–	–	–	–	–	–	–	–	–	–	–	–	–	–	–	–	–	–	–	–	–

Table H.4 Low carbon advance scenario energy balance in 2040

Energy balance for area "Nigeria"

Scenario: Low carbon advance, Year: 2040, Units: Petajoule	Electricity	Natural gas	Gasoline	Kerosene	Diesel	Residual fuel oil	LPG	Crude oil	Coal sub bituminous	Wood	Charcoal	Vegetal wastes	Wind	Solar	Hydro	Nuclear	Lubricants	Biomass	Avgas	CNG	Total
Production	–	715.3	–	–	–	–	–	963.9	0.8	22.3	–	0.2	114.5	236.9	93.6	–	–	570.4	–	–	2,717.9
Imports	–	–	137.5	417.1	–	–	203.8	–	271.1	–	–	–	–	–	–	186.0	–	–	–	–	1,215.5
Exports	–	–	–	–	−39.3	−82.1	–	–	–	–	–	–	–	–	–	–	−108.4	–	−108.4	−108.4	−446.7
Total primary supply	–	715.3	137.5	417.1	−39.3	−82.1	203.8	963.9	271.9	22.3	–	0.2	114.5	236.9	93.6	186.0	−108.4	570.4	−108.4	−108.4	3,486.7
Coal mining	–	–	–	–	–	–	–	–	−0.2	–	–	–	–	–	–	–	–	–	–	–	−0.2
Oil refining	–	–	108.4	108.4	108.4	108.4	108.4	−963.9	–	–	–	–	–	–	–	–	108.4	–	108.4	108.4	−96.4
Charcoal production	–	–	–	–	–	–	–	–	–	−8.2	2.7	–	–	–	–	–	–	–	–	–	−5.5
Electricity generation	663.5	−584.3	–	–	–	–	–	–	−262.3	–	–	–	−114.5	−234.3	−93.6	−186.0	–	−38.6	–	–	−850.1
Transmission and distribution	−33.2	−13.1	–	–	–	–	–	–	–	–	–	–	–	–	–	–	–	–	–	–	−46.3
Total transformation	630.4	−597.4	108.4	108.4	108.4	108.4	108.4	−963.9	−262.5	−8.2	2.7	–	−114.5	−234.3	−93.6	−186.0	108.4	−38.6	108.4	108.4	−998.4
Households	225.0	–	–	525.1	–	–	178.6	–	–	14.1	2.7	0.2	–	2.6	–	–	–	–	–	–	948.3
Industry	188.7	117.9	–	–	–	23.6	–	–	9.4	–	–	–	–	–	–	–	–	132.1	–	–	471.6
Commercial Service	216.7	–	–	–	–	2.7	–	–	–	–	–	–	–	–	–	–	–	234.9	–	–	454.3
Agriculture	–	–	–	0.4	1.4	–	–	–	–	–	–	–	–	–	–	–	–	–	–	–	1.8
Transport	–	–	246.0	–	67.7	–	133.7	–	–	–	–	–	–	–	–	–	–	164.9	–	–	612.2
Total demand	630.4	117.9	246.0	525.5	69.2	26.3	312.2	–	9.4	14.1	2.7	0.2	–	2.6	–	–	–	531.8	–	–	2,488.3
Unmet requirements (Waste)	–	–	–	–	–	–	–	–	–	–	–	–	–	–	–	–	–	–	–	–	–

Table H.5 Green optimistic scenario energy balance in 2040

Energy balance for area "Nigeria"

Scenario: GO, Year: 2040, Units: Petajoule	Electricity	Natural gas	Gasoline	Kerosene	Diesel	Residual fuel oil	LPG	Crude oil	Coal sub bituminous	Wood	Charcoal	Vegetal wastes	Wind	Solar	Hydro	Geothermal	Nuclear	Lubricants	Biomass	Avgas	CNG	Total
Production	–	445.7	–	–	–	–	–	963.9	0.7	29.7	–	0.2	240.4	341.1	120.0	212.2	–	–	471.3	–	–	2,825.2
Imports	–	–	50.2	391.0	–	–	114.9	–	271.0	–	–	–	–	–	–	–	262.1	–	–	–	–	1,089.2
Exports	–	–	–	–	−67.1	−94.4	–	–	–	–	–	–	–	–	–	–	–	−108.4	–	−108.4	−28.4	−406.9
Total primary supply	–	445.7	50.2	391.0	−67.1	−94.4	114.9	963.9	271.7	29.7	–	0.2	240.4	341.1	120.0	212.2	262.1	−108.4	471.3	−108.4	−28.4	3,507.5
Coal mining	–	–	–	–	–	–	–	–	−0.1	–	–	–	–	–	–	–	–	–	–	–	–	−0.1
Oil refining	–	–	108.4	108.4	108.4	108.4	108.4	−963.9	–	–	–	–	–	–	–	–	–	108.4	–	108.4	108.4	−96.4
Charcoal production	–	–	–	–	–	–	–	–	–	−7.4	2.7	–	–	–	–	–	–	–	–	–	–	−4.7
Electricity generation	717.9	−316.5	–	–	–	–	–	–	−263.5	–	–	–	−240.4	−339.1	−120.0	−212.2	−262.1	–	−84.6	–	–	−1,120.7
Transmission and distribution	−28.7	−7.8	–	–	–	–	–	–	–	–	–	–	–	–	–	–	–	–	–	–	–	−36.5
Total transformation	689.1	−324.2	108.4	108.4	108.4	108.4	108.4	−963.9	−263.6	−7.4	2.7	–	−240.4	−339.1	−120.0	−212.2	−262.1	108.4	−84.6	108.4	108.4	−1,258.3
Households	197.3	–	–	499.1	–	–	112.6	–	–	22.2	2.7	0.2	–	1.9	–	–	–	–	–	–	–	836.0
Industry	202.4	121.5	–	–	–	8.1	–	–	8.1	–	–	–	–	–	–	–	–	–	64.8	–	–	404.9
Commercial service	289.4	–	–	–	–	5.9	–	–	–	–	–	–	–	–	–	–	–	–	159.0	–	–	454.3
Agriculture	–	–	–	0.4	1.4	–	–	–	–	–	–	–	–	–	–	–	–	–	–	–	–	1.8
Transport	–	–	158.7	–	39.8	–	110.8	–	–	–	–	–	–	–	–	–	–	–	162.9	–	80.0	552.2
Total Demand	689.1	121.5	158.7	499.4	41.3	14.0	223.3	–	8.1	22.2	2.7	0.2	–	1.9	–	–	–	–	386.7	–	80.0	2,249.2
Unmet requirements (Waste)	–	–	–	–	–	–	–	–	–	–	–	–	–	–	–	–	–	–	–	–	–	–

Appendix I
Results for Least-Cost Electricity Generation

N.V. Emodi, *Energy Policies for Sustainable Development Strategies*,
Frontiers in African Business Research, DOI 10.1007/978-981-10-0974-7

Table I.1 Least-cost electricity generation 2015

Social costs

2015

Branch: Nigeria electricity optimization

Units: Discounted 2010 Cumulative Million U.S. Dollars. Discounted to Year: 2010

Cost categories	Biomass only	CCGT only	CO2 limit	Coal CFB only	Coal IGCC only	Coal steam only	Coal supercritical only	Geothermal only	Large hydro-power only	Nuclear only	Off-shore wind only	On-shore wind only	Optimization	SCGT only	Small hydro-power only	Solar PV only	Solar thermal only	Total
Transformation capital	1.0	0.4	1.9	0.6	0.8	0.6	0.6	0.6	0.5	1.1	2.9	0.9	1.9	0.4	1.1	2.9	1.7	19.9
Transformation fixed O&M	0.5	0.2	0.2	0.4	0.4	0.1	0.3	0.4	0.1	1.1	0.5	0.3	0.2	0.1	0.1	0.1	0.6	5.6
Transformation variable O&M	573.7	367.9	–	362.7	580.4	281.5	507.8	441.3	27.7	367.8	–	–	–	220.7	–	–	–	3,731.5
Fuel production	–	3,466.5	–	3,307.5	3,087.0	4,491.4	3,157.2	–	–	–	–	–	–	5,024.7	–	–	–	22,534.3
Environmental externalities	543.7	494.2	–	1,417.3	1,862.2	2,709.4	1,904.5	–	–	–	–	–	–	716.4	–	–	–	9,647.8
Total	1,119.0	4,329.2	2.1	5,088.6	5,530.8	7,482.9	5,570.4	442.4	28.4	369.9	3.4	1.2	2.1	5,962.3	1.2	3.0	2.2	35,939.1

Table I.2 Least-cost electricity generation 2020

Social costs

2020

Branch: Nigeria electricity optimization

Units: Discounted 2010 Cumulative Million U.S. Dollars. Discounted to Year: 2010

Cost categories	Biomass only	CCGT only	CO2 limit	Coal CFB only	Coal IGCC only	Coal steam only	Coal supercritical only	Geothermal only	Large hydro-power only	Nuclear only	Off-shore wind only	On-shore wind only	Optimization	SCGT only	Small hydro-power only	Solar PV only	Solar thermal only	Total
Transformation capital	3.2	1.3	6.0	1.9	2.6	1.8	2.0	1.9	1.7	3.3	8.9	2.8	6.0	1.2	3.3	8.9	5.2	62.1
Transformation fixed O&M	1.6	0.5	0.7	1.3	1.1	0.3	0.9	1.3	0.4	3.3	1.5	0.8	0.7	0.4	0.4	0.4	1.7	17.3
Transformation variable O&M	1,790.5	1,152.7	–	1,132.9	1,812.7	878.1	1,586.1	1,382.1	86.2	1,151.8	–	–	–	691.1	–	–	–	11,664.2
Fuel production	–	10,861.4	–	10,330.7	9,641.9	14,012.9	9,861.1	–	–	–	–	–	–	15,735.9	–	–	–	70,443.9
Environmental externalities	1,696.8	1,548.6	–	4,426.9	5,816.4	8,453.2	5,948.6	–	–	–	–	–	–	2,243.5	–	–	–	30,134.0
Total	3,492.1	13,564.5	6.7	15,893.7	17,274.8	23,346.3	17,398.6	1,385.3	88.3	1,158.3	10.5	3.7	6.7	18,672.0	3.7	9.3	7.0	112,321.5

Table I.3 Least-cost electricity generation 2025

Social costs

2025

Branch: Nigeria electricity optimization

Units: Discounted 2010 Cumulative Million U.S. Dollars. Discounted to Year: 2010

Cost categories	Biomass only	CCGT only	CO2 limit	Coal CFB only	Coal IGCC only	Coal steam only	Coal supercritical only	Geothermal only	Large hydro-power only	Nuclear only	Off-shore wind only	On-shore wind only	Optimization	SCGT only	Small hydro-power only	Solar PV only	Solar thermal only	Total
Transformation capital	6.0	2.4	11.3	3.5	4.8	3.4	3.7	3.5	3.1	6.1	16.7	5.3	11.3	2.3	6.2	16.6	9.8	115.8
Transformation fixed O&M	3.1	1.0	1.3	2.4	2.1	0.5	1.6	2.4	0.8	6.1	2.8	1.5	1.3	0.7	0.7	0.7	3.2	32.2
Transformation variable O&M	3,344.6	2,156.8	–	2,116.7	3,386.7	1,640.3	2,963.4	2,585.0	160.7	2,154.2	–	–	–	1,292.5	–	–	–	21,801.0
Fuel production	–	20,321.6	–	19,301.3	18,014.5	26,175.2	18,423.9	–	–	–	–	–	–	29,431.2	–	–	–	131,667.6
Environmental externalities	3,169.7	2,897.3	–	8,271.0	10,867.1	15,790.0	11,114.1	–	–	–	–	–	–	4,196.1	–	–	–	56,305.3
Total	6,523.4	25,379.0	12.6	29,694.8	32,275.2	43,609.3	32,506.6	2,591.0	164.6	2,166.4	19.5	6.8	12.6	34,922.7	7.0	17.3	13.0	209,921.9

Table I.4 Least-cost electricity generation 2030

Social costs

2030

Branch: Nigeria electricity optimization

Units: Discounted 2010 Cumulative Million U.S. Dollars. Discounted to Year: 2010

Cost categories	Biomass only	CCGT only	CO2 limit	Coal CFB only	Coal IGCC only	Coal steam only	Coal supercritical only	Geothermal only	Large hydro-power only	Nuclear only	Off-shore wind only	On-shore wind only	Optimization	SCGT only	Small hydro-power only	Solar PV only	Solar thermal only	Total
Transformation capital	9.0	3.6	17.0	5.2	7.2	5.1	5.5	5.2	4.6	9.2	25.1	8.0	17.0	3.4	9.4	25.0	14.7	174.2
Transformation fixed O&M	4.6	1.5	2.0	3.6	3.1	0.7	2.4	3.7	1.2	9.2	4.2	2.3	2.0	1.0	1.1	1.0	4.9	48.5
Transformation variable O&M	5,033.3	3,248.3	–	3,185.6	5,097.0	2,468.4	4,459.8	3,892.4	241.7	3,243.7	–	–	–	1,946.2	–	–	–	32,816.4
Fuel production	–	30,606.3	–	29,048.1	27,111.5	39,390.0	27,727.7	–	–	–	–	–	–	44,315.6	–	–	–	198,199.2
Environmental externalities	4,770.0	4,363.6	–	12,447.7	16,354.8	23,761.7	16,726.5	–	–	–	–	–	–	6,318.2	–	–	–	84,742.7
Total	9,816.9	38,223.2	19.0	44,690.2	48,573.6	65,626.1	48,921.9	3,901.3	247.5	3,262.1	29.3	10.3	19.0	52,584.5	10.5	26.0	19.6	315,981.0

Table I.5 Least-cost electricity generation 2035

Social costs

2035

Branch: Nigeria electricity optimization

Units: Discounted 2010 Cumulative Million U.S. Dollars. Discounted to Year: 2010

Cost categories	Biomass only	CCGT only	CO2 limit	Coal CFB only	Coal IGCC only	Coal steam only	Coal supercritical only	Geothermal only	Large hydro-power only	Nuclear only	Off-shore wind only	On-shore wind only	Optimization	SCGT only	Small hydro-power only	Solar PV only	Solar thermal only	Total
Transformation capital	12.0	4.8	22.7	7.0	9.6	6.8	7.3	7.0	6.2	12.3	33.5	10.7	22.7	4.5	12.5	33.4	19.6	232.7
Transformation fixed O&M	6.1	2.0	2.7	4.8	4.2	1.0	3.2	4.9	1.6	12.3	5.7	3.1	2.7	1.4	1.5	1.4	6.5	64.8
Transformation variable O&M	6,726.1	4,342.4	–	4,257.2	6,811.4	3,298.6	5,960.0	5,202.9	322.9	4,335.8	–	–	–	2,601.5	–	–	–	43,858.7
Fuel production	–	40,915.2	–	38,819.0	36,231.1	52,637.6	37,054.5	–	–	–	–	–	–	59,236.1	–	–	–	264,893.5
Environmental externalities	6,374.3	5,833.4	–	16,634.8	21,856.1	31,753.2	22,352.8	–	–	–	–	–	–	8,445.5	–	–	–	113,250.2
Total	13,118.6	51,097.7	25.3	59,722.7	64,912.4	87,697.2	65,377.9	5,214.8	330.7	4,360.4	39.2	13.7	25.3	70,289.0	14.0	34.7	26.2	422,299.9

Table I.6 Least-cost electricity generation 2040

Social costs

2040

Branch: Nigeria electricity optimization

Units: Discounted 2010 Cumulative Million U.S. Dollars. Discounted to Year: 2010

Cost categories	Biomass only	CCGT only	CO2 limit	Coal CFB only	Coal IGCC only	Coal steam only	Coal supercritical only	Geothermal only	Large hydro-power only	Nuclear only	Off-shore wind only	On-shore wind only	Optimization	SCGT only	Small hydro-power only	Solar PV only	Solar thermal only	Total
Transformation capital	14.9	5.9	28.1	8.7	12.0	8.5	9.1	8.7	7.7	15.3	41.5	13.2	28.1	5.6	15.5	41.4	24.3	288.6
Transformation fixed O&M	7.6	2.4	3.3	5.9	5.2	1.2	4.0	6.1	2.0	15.2	7.0	3.8	3.3	1.7	1.8	1.7	8.1	80.3
Transformation variable O&M	8,340.8	5,386.8	–	5,279.2	8,446.8	4,090.5	7,390.9	6,453.4	400.2	5,377.8	–	–	–	3,226.7	–	–	–	54,393.2
Fuel production	–	50,755.9	–	48,138.8	44,929.6	65,273.8	45,950.7	–	–	–	–	–	–	73,473.1	–	–	–	328,521.9
Environmental externalities	7,904.6	7,236.4	–	20,628.6	27,103.4	39,375.9	27,719.4	–	–	–	–	–	–	10,475.3	–	–	–	140,443.5
Total	16,267.9	63,387.5	31.4	74,061.2	80,496.9	108,749.8	81,074.1	6,468.1	409.9	5,408.3	48.6	17.0	31.4	87,182.5	17.4	43.0	32.4	523,727.5

Table I.7 Power plant specific GHGs

	Biomass only			CCGT only			Coal CFB only			Coal IGCC only		
	2020	2030	2040	2020	2030	2040	2020	2030	2040	2020	2030	2040
Carbon dioxide non-biogenic	0	0	0	20.77363	41.54727	62.3209	46.7535	93.21135	139.9648	43.6366	86.99726	130.6339
Methane	0.164804	0.328578	0.493382	0.007821	0.015641	0.023462	0.010598	0.021129	0.031726	0.009891	0.01972	0.029611
Nitrous oxide	0.324377	0.646725	0.971102	0.011545	0.02309	0.034634	15.01859	29.9422	44.96079	0.20442	0.407547	0.611966
Total	0.489181	0.975303	1.464484	20.793	41.586	62.379	61.78268	123.1747	184.9574	43.85091	87.42453	131.2754

Table I.8 Power plants specific GHGs

	Coal steam only			Coal supercritical only			SCGT only		
	2020	2030	2040	2020	2030	2040	2020	2030	2040
Carbon dioxide non-biogenic	63.38541	126.3813	189.7667	44.62834	88.97447	133.6028	30.0966	60.09453	90.19113
Methane	0.014368	0.028647	0.043015	0.010116	0.020168	0.030284	0.011331	0.022624	0.033954
Nitrous oxide	0.296935	0.592045	0.888979	0.209066	0.416809	0.625875	0.016726	0.033397	0.050123
Total	63.69671	127.002	190.6987	44.84752	89.41145	134.259	30.12466	60.15055	90.27521

References

Adenikinju AF (2003) Electric infrastructure failures in Nigeria: a survey-based analysis of the costs and adjustment responses. Energy Policy 31(14):1519–1530

Akpan US, Isihak SR, Udoakah YON (2013) Electricity access in Nigeria: viability of off-grid photovoltaic system. In: AFRICON, 2013, pp 1–8. IEEE

Anumaka MC (2012) Analysis of technical losses in electrical power system (Nigerian 330 kV network as a case study). Department Of Electrical Electronic Engineering, Faculty of Engineering, Imo State University, Owerri, Imo State, Nigeria Email: engranumakamc@ yahoo. com

Babatunde S, Akintayo (2012) World Bank vehicle population survey in Nigeria. Nigerian Institute of Transport Technology (NITT), Department of Professional Transport Studies, Zaria, Nigeria

Breeze P (2014) Power generation technologies. Newnes

Branker K, Pathak MJM, Pearce JM (2011) A review of solar photovoltaic levelized cost of electricity. Renew Sustain Energy Rev 15(9):4470–4482

Cervigni R, Dvorak I, Rogers JA (eds) (2013) Assessing low-carbon development in Nigeria: an analysis of four sectors. World Bank Publications

Emodi NV, Boo KJ (2015) Decomposition analysis of CO_2 emissions from electricity generation in Nigeria. Int J Energy Econ Policy 5(2):565–573

Energy Commission of Nigeria (ECN), (FGN) Federal Government of Nigeria (2013) National Energy Policy (NEP). Plot 701C, Central Business District, Abuja, Nigeria

Energy Commission of Nigeria (ECN), (FMST) Federal Ministry of Science and Technology, (FGN) Federal Government of Nigeria (2014a) National Energy Master Plan (NEMP). Plot 701C, Central Business District, Abuja, Nigeria

Energy Commission of Nigeria (ECN), (FMST) Federal Ministry of Science and Technology (2014b) National Renewable Energy and Energy Efficiency Policy (NREEEP). Plot 701C, Central Business District, Abuja, Nigeria

European Environmental Agency (EEA) (2015) Environmental tax reform: increasing individual incomes and boosting innovation. www.eea.europa.eu/highlights/environmental-tax-reform-increasing-individual. Accessed 5 Sept 2015

Fadel M El, Chedid R, Zeinati M, Hmaidan W (2003) Mitigating energy-related GHG emissions through renewable energy. Renewable Energy 28(8):1257–1276

Federal Republic of Nigeria (FGN) (1999) Combating desertification and mitigating the effects of drought in Nigeria. National Report on the Implementation of the United Nations Convention to Combat Desertification in those Countries Experiencing Serious Drought and/or Desertification, Particularly in Africa (CCD). Submission at the Third Session of the Conference of the Parties, Recife, Brazil, November 1999. www.unccd-prais.com/.../4f8e506b-d10d-450b-8a1e-a0fa014a4a65. Accessed 6 April 2015

Gallivan F (2013) Energy savings strategies for transit agencies, vol 106. Transportation Research Board

Gujba H, Mulugetta Y, Azapagic A (2015) The household cooking sector in Nigeria: environmental and economic sustainability assessment. Resources 4(2):412–433

Howells M, Rogner H, Strachan N, Heaps C, Huntington H, Kypreos S, Hughes A, Silveira S, DeCarolis, Bazilian M, Roehrl A (2011). OSeMOSYS: the open source energy modeling system: an introduction to its ethos, structure and development. Energy Policy 39(10):5850–5870

Ibitoye FI (2013) The millennium development goals and household energy requirements in Nigeria. SpringerPlus 2(1):1–9

Irimiya Y, Humphery IA, Aondover II (2013) Assessment of energy use pattern in residential buildings of Kano and Kaduna Northern Nigeria. Am J Eng Res 2:271–275

Ibitoye FI (2013) The millennium development goals and household energy requirements in Nigeria. SpringerPlus 2(1):1–9

Khatib H (2010) Review of OECD study into "Projected costs of generating electricity—2010 Edition". Energy Policy 38(10):5403–5408

Kurowska E, Krzysztof S (2010) Geothermal exploration in Nigeria. In: Proceedings world geothermal congress

Linnér BO, Selin H (2013) The United Nations conference on sustainable development: forty years in the making. Environ Plann C: Govern Policy 31(6):971–987

Muhammad U (2012) Rural solar electrification in Nigeria: renewable energy potentials and distribution for rural development. In: SOLAR2012_0332

Mustonen SM (2010) Rural energy survey and scenario analysis of village energy consumption: a case study in Lao People's Democratic Republic. Energy Policy 38(2):1040–1048

National Population Commission (2014) Nigeria demographic and health survey 2013. National Population Commission [Nigeria]. http://www.population.gov.ng/images/ndhs_data/ndhs_2013/2013_ndhs_final_report.pdf. Accessed 12 Aug 2014

Rahman MSU (2009) Fuel consumption of transport sector: how the people of Dhaka city will be moving in the future?. Act! Innovate! Deliver! Reducing energy demand sustainably. In: Proceedings of the ECEEE summer study, pp 409–1415

Renewable Energy Programme (RUWES), Federal Ministry of Environment of Nigeria (2015) Submission to the United Nations framework convention on climate change (UNFCCC) on the Lima programme on gender. http://www4.unfccc.int/submissions/Lists/OSPSubmissionUpload/39_122_130688171405397115-Nigeria%20gender.pdf. Accessed 18 Dec 2015

Salvatore J (2013) World Energy perspective: cost of energy technologies. World Energy Council, London

Sambo AS (2008) Matching electricity supply with demand in Nigeria. In: International Association for Energy Economics, (Fourth Quarter), pp 32–38

Sims RE, Mabee W, Saddler JN, Taylor M (2010a) An overview of second generation biofuel technologies. Bioresour Technol 101(6):1570–1580

Sims RE, Mabee W, Saddler JN, Taylor M (2010b) An overview of second generation biofuel technologies. Bioresour Technol 101(6):1570–1580

Stefansson V (2002) Investment cost for geothermal power plants. Geothermics 31(2):263–272

Tallapragada PVSN (2009) Nigeria's electricity sector-electricity and gas pricing barriers. In: International Association for Energy Economics, pp 29–34

Taoa Z, Liu Z, Zhao Changxin Z (2011) Research on the prospects of low-carbon economic development in China based on LEAP model. Energy Procedia 5:695–699

Taylor P, D'ortigue OL, Trudeau N, Francoeur M (2008) Energy efficiency indicators for public electricity production from fossil fuels IEA information paper in support of the G8 plan of action

The Federal Government of Nigeria—National Planning Commission (FGN-NPC) (2010) Nigeria Vision 20:2020. In: The 1st NV20:2020 medium term implementation plan (2010–2013), vol 1: the vision and development priorities. www.preventionweb.net/files/14632_1stnipeditedversionvol1.pdf. Accessed 25 July 2015

Tidball R, Bluestein J, Rodriguez N, Knoke S (2010) Cost and performance assumptions for modeling electricity generation technologies. Contract 303:275–3000

UNFCCC (2008) Module 4: measures to mitigate climate change, resource guide for preparing the national communications of non-Annex I parties. Bonn, Germany

Wiltsee G (2000) Lessons learned from existing biomass power plants (No. NREL/SR-570-26946). National Renewable Energy Lab., Golden, CO (US)

Zarrouk SJ, Moon H (2014) Efficiency of geothermal power plants: A worldwide review. Geothermics 51:142–153

Index

N.V. Emodi, *Energy Policies for Sustainable Development Strategies*,
Frontiers in African Business Research, DOI 10.1007/978-981-10-0974-7

MIX
Papier aus verantwortungsvollen Quellen
Paper from responsible sources
FSC® C105338

FSC
www.fsc.org

If you have any concerns about our products,
you can contact us on
ProductSafety@springernature.com

In case Publisher is established outside the EU,
the EU authorized representative is:
Springer Nature Customer Service Center GmbH
Europaplatz 3, 69115 Heidelberg, Germany

Printed by Libri Plureos GmbH
in Hamburg, Germany